THE
MULLAPERIYAR
WATER WAR

After completing his graduation in chemistry and post-graduation in computer application, **Pradeep Damodaran** pursued a successful career in the software industry for nine years before quitting IT to take up journalism. He is presently chief of bureau, *Deccan Chronicle*, Chennai.

THE MULLAPERIYAR

WATER WAR

THE DAM THAT DIVIDED
TWO STATES

PRADEEP DAMODARAN

RUPA

Published by
Rupa Publications India Pvt. Ltd 2014
7/16, Ansari Road, Daryaganj
New Delhi 110002

Sales centres:
Allahabad Bengaluru Chennai
Hyderabad Jaipur Kathmandu
Kolkata Mumbai

ISBN: 978-81-291-3560-5

First impression 2014

10 9 8 7 6 5 4 3 2 1

The moral right of the author has been asserted.

Printed at Parksons Graphics Pvt. Ltd, Mumbai

Contents

Introduction

Deep down in south Tamil Nadu (TN), the southern-most part of the country, regional politicians cite three historical injustices done to the Tamil race whenever they want to rouse public sentiment in their favour. These include the mass genocide of Tamils in Sri Lanka, the injustice done to TN in sharing the Cauvery water, and the continued ignorance to the plight of Tamils in the Mullaperiyar dam controversy.

Whenever a public speaker rakes up any of these three issues in south TN, especially in and around the bustling city of Madurai, he is certain to get a loud applause and a bunch of youths who are willing to do anything to protect the Tamil cause, even if it means going to war with one's own countrymen.

The fact that the water-sharing dispute between neighbouring states and the safety concerns of a century-old dam invokes hatred and anger, which is as intense as one's ire against killing of thousands of people belonging to the same race, shows the importance of water disputes in a country as ethnically diverse as India and its potential to explode into a full-fledged riot.

The situation on the other side of the Western Ghats in central and south Kerala is no different. To the Malayalis, the Tamils are a thankless race. Talk to any Malayali in Idukki district and the same argument repeats: 'They take away our water for more than a hundred years and then they claim to be historically cheated. Despite benefitting from our waters, the Tamils are not concerned about the lives of thousands of Malayalis who could get killed if the old dam breaks,' is the thought in the mind of every Malayali in Idukki district.

While there has been excellent cooperation between Tamils and Malayalis for centuries, when the topic of Mullaperiyar or Mullaiperiyar (depending on which side of the Western Ghats you are in) is taken, the bonhomie vanishes in a jiffy.

Interstate river water disputes might not carry with them a threat of escalating into an all out war but could be equally, if not more, damaging for a nation. Especially in a country such as India, where individual states have their own cultural and linguistic identities, an interstate river water conflict is sufficient to give vent to pent up hatred for the neighbouring state, as we have seen time and again.

The first major legislation that was brought in to provide a set of guidelines for adjudicating and resolving these disputes came in the form of the Inter State River Water Disputes Act of 1956. As per the act, any state government that feels it has a dispute with another state over sharing a river could approach the central government to constitute a tribunal to resolve the dispute. This act was mainly introduced to protect the interests of the downstream states who could be deprived of their right to a share of the river by activities of the upstream states. The only time when an upstream state could invoke the act is when the reservoir formed out of a dam built by the downstream state extends into the neighbouring territory. This act was later modified over the years to cater to the changing water needs of the country.

According to the Ministry of Water Resources, there are presently five interstate water disputes and tribunals. They are Ravi and Beas Waters Tribunal, Vansadhara River Water Dispute, Mahadayi/Mandovi River Dispute, Krishna River Water Disputes Tribunal and Cauvery River Water Disputes.

Most of these river water disputes have been raging for several decades and continue to hamper development activities in the states. Be it the Ravi–Beas water dispute between Punjab and Haryana or the Cauvery water dispute between TN and Karnataka or any other water-sharing controversies they soon get mixed up with local politics and gain communal identities, and over a period of time turn into complex jigsaw puzzles that nobody dares to solve.

International water management experts who have attempted to study water conflicts in India blame the lack of a proper set of guidelines on par with international standards as the reason for such an inordinate delay in resolving these disputes, and have urged for legal reforms before the situation gets out of control.

But, in a country like India, tribunal verdicts could have catastrophic outcomes, as happened during the winter of 1991 in Bengaluru following the gazette notification of the verdict of Cauvery River Water Disputes Tribunal.

*

In December 1991, the Government of India gazetted the interim order of the Cauvery River Water Disputes Tribunal ensuring 205 tmc of water every year for TN. Two days later, one of the most violent riots broke out in Karnataka in which over eighteen persons were killed, hundreds of Tamil women were molested, and thousands of Tamilians who had been living in the metropolitan city for generations were lynched and forced to flee the state, according to reports published at that time.

Veteran journalists in the region point out that the Congress-led state government remained mute as angry mobs of Kannada chauvinists raided poor Tamil neighbourhoods armed with knives and wooden logs, beating up Tamilians wherever they met. Only after two days of violence and more than a dozen deaths were reported did the police step in and try to control the unruly mobs. Soon, violence spread to other parts of the state and at least two lakh Tamil families are reported to have fled the state, leaving behind their homes and property never to return back to Karnataka.

The riots that lasted for two weeks and caused irreparable damage to the relationship between the two states was the fallout of a decades-long dispute in sharing the waters of the Cauvery River that is the lifeline for the farmers in both Karnataka and TN.

Unlike North Indian rivers that mostly originate from the Himalayas and are fed by glaciers, all the rivers in South India are rain fed and originate mostly from the Western Ghats. The Cauvery rises in the Brahmagiri range of the Western Ghats in Coorg district of Karnataka at an elevation of about 1,340 metres. It is joined by many smaller tributaries besides the Kabini River that flows eastward from Kerala and joins the Cauvery before it flows into TN.

At the point of its entrance into TN is the Bhavanisagar dam in Mettur beneath which three more rivers, Bhavani, Noyyal and Amaravathi, join the Cauvery as it flows into the fertile plains of Thanjavur district. Presently, four different states including Karnataka, Kerala, TN and Puducherry share riparian rights for the river waters, and the dispute between TN and Karnataka farmers in sharing the Cauvery water is almost a hundred years old.

More than six years after a final verdict was pronounced, the decision of the Cauvery Water Disputes Tribunal was eventually notified in February 2013. TN has won a sizeable share of the Cauvery water, but it still remains only on paper as water from the Cauvery is yet to flow down south in bountiful amounts, as ordered by the courts.

*

The Mullaperiyar dam row between TN and Kerala, which has been persistently flaring for the last hundred years and more, is not just another interstate river water dispute. It is much more complex and, at the same time, a fascinating story of a dam that was considered a technological marvel at the time it was constructed. The engineer who built the dam is still revered as God in the five districts of TN that benefit from the dam.

Among the many awesome details of the dam is its achievement of performing a successful interbasin river transfer, probably the first of its kind in the world. The dam built on the Periyar River diverts a west flowing river to the east, thereby transforming an arid part of the country into one of the most fertile landscapes.

Mullaperiyar dam and its reservoir are located entirely in the state of Kerala but serve the people of TN. For more than a hundred years, despite all the conflicts surrounding its safety, the dam has stood as a symbol of interstate cooperation with mutually beneficial results. While Tamil farmers get adequate water to cultivate crops, a vast majority of the harvest finds a cheap market in Kerala, where there is not sufficient land for agriculture owing to the mountainous terrain.

As the decades passed by, the dam aged gracefully. Over all these years the needs and expectations of people living on both sides of the Mullaperiyar changed with the changing times, and water became a lot more precious than it was a hundred years ago.

The present cause of conflict is an outcome of the 119 years of social change that took place in TN and Kerala. Today, the dam is a source of life for millions of people, and remains a mute spectator to all the controversies surrounding it.

Several panels of judges and empowered committees have been formed by the central government to solve the Mullaperiyar dam row in the past. But they have all been mere exercises in futility.

To solve a dispute regarding a dam as old and marvellous as the Mullaperiyar, one has to look at it with the same passion and objectivity that the proponents of the dam had when they mooted the idea 150 years ago.

*

1
Flashpoint

Exactly twenty years after the Cauvery water-related violence was unleashed, another water-related riot broke out down South in the mountainous border between TN and Kerala in December 2011. This time, the dispute was concerning the safety of the 119-year-old Mullaperiyar dam and its disputed storage level.

Thousands of vehicles were lynched, hundreds of homes were looted and although no one was killed, several dozen persons were injured as angry mobs of Tamils and Keralites went on a rampage against the other in their respective states. As with all other riots, there was no account of the violence unleashed, the vehicles that were damaged and the homes that were looted.

At the heart of this conflict was the safety and storage capacity of the 119-year-old Mullaperiyar dam constructed in the middle of a tiger reserve by diverting a large river against its natural flow. While the giant structure is in present-day Kerala and continues to block a mighty river for more than a century, the beneficiaries of the project live in TN—a cause for bitterness and distrust between the two states and its people.

*

Sudhakar Reddy (34), from Hyderabad, had been holed up inside a tiny makeshift shelter, built by a local Ayyappa Seva Trust along the Theni–Kumily Highway barely a few kilometres away from Cumbum, for the last three days with nine other persons from his hometown. The independent software consultant, who looked like a sanyasi with a long beard wearing just a saffron dhoti with beads of different sizes hanging down his neck, had been on a padayatra to the famous Sabarimala Ayyappa Shrine in Kerala from Hyderabad, along with a group of devotees, when his journey took an unexpected turn.

'I did not expect the last leg of our journey to end up like this,' he told me, as he sat on a coir mat shared by two others, also from Andhra Pradesh. 'We have been told by the police not to go any further as they suspect that we could be mistaken for Tamilians and attacked.'

At least a dozen other devotees, who had been on the padayatra from different parts of the country to Sabarimala, were stuck along with Sudhakar and his group for the last several days, and were put up in many makeshift shelters that came up along the Theni–Cumbum Road. Local volunteers from the Seva Trust cooked and served free meals to the pilgrims and ensured that they had a roof over their head until the situation improved.

I met Sudhakar at around 9.30 p.m. on 6 December 2011 on my way to Cumbum to report on the violence that had been unleashed against Malayalis over the last few days. Upon seeing a crowd near the pandal surrounded by a police battalion, I got out of my cab to enquire.

'The police are advising us to go back and come next year. But I have not come so far to return home without getting a darshan and performing my pooja,' he said.

Many others, who had been stuck outside Cumbum like Sudhakar, decided to cut short their journey and performed poojas at a local temple near Theni and returned home. Several even chose to go to the Murugan temple in Palani, another famous shrine in this region, and went back. Sudhakar and his group, however, hoped for the situation to return to normal and finish their trip.

After interviewing Sudhakar and his group and clicking a few pictures, I went out to talk to the police personnel posted outside on the situation and whether it was safe to go to Kumily so late in the night.

Just then, a private luxury bus stopped ahead of us with its windowpanes broken and the front headlights smashed to pieces. The driver, Karivaratharaj, who was still shivering, jumped out of the driver's seat and rushed to the police.

'They are trying to kill us. As I drove near Kumily, a group hurled huge stones at us. The glass window was broken to pieces and I almost lost control of the vehicle. The rock pieces were so big that I had to hide my face to avoid getting killed,' he said. Shocked passengers, all Sabarimala pilgrims returning after their annual visit, got out of the bus and recounted the experience in horrific detail. 'They hurled stones and stopped the vehicle. Then a mob rushed to the bus with wooden logs and hit the side frame of the bus hurling abuses at us,' said a young man, seething with anger.

Soon, a crowd gathered around the bus and listened to the tale. As the numbers multiplied, they started raising slogans against the Kerala chief minister and Keralites. 'We will teach them a lesson. Do not allow any KL-registration vehicles to pass this road,' a member of the crowd told the police gathered there.

As I took out my camera to shoot pictures of Karivaratharaj and the bus, one of the men grabbed my camera and asked me for my identity card. I dug it out of my haversack bag and showed him. The crowd was still furious. A sub inspector intervened, checked my identity card and asked me to leave the place. 'These people are very agitated and drunk, don't entertain them, sir. You keep going and don't stop your car on the way,' he said.

I reached my hotel room in Cumbum town that had already been booked in advance, at around 11.00 p.m. and started filing a report on the bus attack. My cab driver Sarvanan, who was a native of Cumbum and had driven me all the way from Coimbatore, had gone to pick up dinner for the two of us. Even before I could finish the report and send it, Sarvanan rushed to my room and informed me that a bus bound for Kerala was being burnt along the Cumbummettu Road and that other media persons were rushing to the spot.

Within a few minutes, we were back in the car heading towards Cumbummettu around 5 kilometres from the hotel. The road was desolate except for groups of armed reserve police personnel who were on foot patrol.

When we reached Cumbummettu, a police team blocked our vehicle and asked us to go back as no one was allowed beyond that point. I tried to argue that I was a journalist and had already spoken to the superintendent of police (SP) and obtained his permission (a lie). The police inspector, a woman, refused to buy it and sent me back.

As we turned back, I noticed some kind of vehicle burning at a distance beyond the police barricade. I zoomed my camera, clicked a few pictures of the burning vehicle and returned to the hotel.

*

When I woke up the next morning at Cumbum, my mobile phone was already flooded with messages from my boss. '500 Tamil women molested in Kerala...Bus burnt down and passengers beaten...Thousands flee to their hometowns...Looting and arson rampant...' The list went on. These were headlines of stories that appeared in other papers that I had missed.

In the next thirty minutes, I was back on the road and heading to the local SP's office for a press meet. When I reached the office, they said the press meet was cancelled as more violence had broken out in Bodhimettu, one of the three roads that led to Kerala from Theni district.

The biggest problem with covering such events is that the reporter is always late. When you reach a spot, they say it just got over and ask you to rush elsewhere. To be honest, I did not witness any real violence that day as I was always late despite dashing through the dingy streets at breakneck speed. Just groups of protestors burning effigies of the Kerala chief minister and organizing protest marches were all I could cover.

While covering riots and violence in otherwise peaceful regions is nothing compared to covering a war or reporting from conflict zones such as Kashmir, it could still be a scary affair for print journalists. Unlike television media reporters, who usually come with a cameraman and a few assistants besides an OB van and driver, print journalists are often sent alone having to double up as photographers. Although one could rest assured that he or she would not be killed while covering a riot, a couple of hard blows on the face while taking a picture or the possibility of getting in the way of a flying piece of rock are very real and unnerving. Angry mobsters take special pleasure in breaking expensive cameras to pieces for they know that they can get away with it.

After fixing a deal with a couple of local sources, of course for a fee, I went around town to interview the 500 women who claimed to have been molested, accompanied by two locals, Ashok and Kumar. I was told that most of the victims were from the Uthamapuram North neighbourhood and rushed there. Of the 500 women who were allegedly molested, only Virumayi (70), from Uthamapuram North, agreed to speak and greeted us with a smile. Despite her pleasant smile and animated expressions, Virumayi actually looked like she was at least fifteen years older.

Kumar and Ashok who accompanied me pointed at her and said that she would tell me what exactly happened. 'The younger women will not be willing to talk,' Kumar said.

Virumayi had been going to the cardamom estates in Kerala to work as an estate labourer for the last two decades. 'In all these years I have never seen anything like this. They stopped our van and asked all the women to get out. Then they broke down the vehicle and burnt it,' she said.

'But did they molest 500 women?'

'They pulled our saris and used bad words. Most of the men were very angry and tried to chase us into the fields. Since I was old, I managed to walk back,' she said. The elderly woman swore that she had not seen anything like that all her life and would never go back to Kerala.

The day before I reached Cumbum, where the incident allegedly took place, Kerala Chief Minister Oomen Chandy had sat on a one-day hunger strike to press for a new dam and to express solidarity with the people of Idukki district who were on a relay protest demanding safety for their lives. Congress supporters, who were eagerly waiting to show their support for the leader, unleashed heavy violence and broke at least a dozen Tata Sumo cars that were mostly used to transport estate workers from Cumbum and surrounding areas to Idukki district and back, leaving hundreds of such female workers stranded. While many managed to return home later during the day, I was told that several among them were still holed up.

'We went to the police immediately but they refused to help and they chased us away saying "Poda pandi, go and complain to your police",' Kumar told me as we escorted Virumayi back to her hut.

I asked Saravanan to take the cab straight to the Cumbummettu police station that was in Kerala jurisdiction at the interstate border to find out why not even a single FIR was registered. Despite the heavy police presence on both sides of the border, we managed to sneak through using my press card and reached the Cumbummettu police station.

At least three different jeeps were charred and lying next to the police station. Glass pieces from shattered windowpanes were strewn all over the road. When we reached the police station, I asked my driver and local contacts to wait outside and went in.

'The inspector is not here. Come back later,' said a constable.

I told him that I was a reporter and wanted to talk to an officer. After waiting ten more minutes, I was taken to a sub inspector who asked for my identity card.

'You are from Tamil Nadu?'

'Yes.'

'Why don't you go and ask your police why they are not registering cases?' he said.

*

The long and winding Cumbummettu Road, which meanders amid the lush green vineyards and rolling hills of the Western Ghats before making a steep hike to the top of a craggy hill through eighteen hairpin bends to reach God's Own Country, hardly resembles a border. While there are several routes to reach

Kerala from TN, this one is the shortest from any of the southern districts, used by hundreds of buses, vans and cars every day.

For the last twenty years, Roy (54), a Malayali hotelier who had settled down in Cumbum, has been driving along this scenic road to meet his relatives on the 'God's Own' side of the world. But when I first met him at the interstate border with a dozen other relatives and friends of his, it did seem like he had just escaped from a terrorists' base in Pakistan and reached Indian borders. The border had been sealed; at least half a dozen police check posts had been set up along the Cumbummettu Road and no one was allowed to cross over to the other state unless he or she was going home.

Roy and his family have been living in Cumbum, the scenic yet noisy little town surrounded by vineyards in Theni district bordering Kerala, for decades. He owns a popular eatery situated in the heart of town which is famous across the region for its spicy beef curry and cuppa (a dish made by steaming tapioca and cooking it with spices).

In his early thirties, he had moved to TN seeking a better fortune after trying several odd jobs back in his native village, and later started the hotel with a paltry sum as investment. As his business grew, Roy married Bini, also from back home in Kerala, and the couple has two teenage daughters who are studying at a private convent near Theni.

Being a shrewd businessman, he never closed his hotel for more than a day during the last two decades. Since early December 2011, the establishment has remained locked and been repeatedly vandalized week after week by his former friends and clients.

'I have never felt insecure living in Cumbum for all these years,' he said. 'I have been here for so long that I know most of the people in Cumbum. Almost 90 per cent of this town's population must have eaten from my hotel and I know most of my customers by their first name.'

I met Roy three days after the violence first broke out in TN and Kerala over the Mullaperiyar dam controversy exactly at the interstate border adjoining the Cumbummettu police station. A deputy superintendent of police (DSP) and a team of TN policemen had just rescued the hotelier, his wife Bini, their two daughters, a servant, and half a dozen other Malayalis from their homes in Cumbum and driven them in a police van whose windowpanes were protected by iron grills to the interstate border 'to safety'.

The proud DSP pointed his fingers at the group who were now huddled near him like refugees and said, 'See, sir, we have safely rescued all these Malayalis and have brought them home. Please write about these things in the newspaper. Tamil Nadu police is taking all necessary steps to protect Keralites who are holed up here. By tonight, we would have rescued all of them,' he said.

The authoritative tone of his voice suggested the satisfaction of accomplishment of a feat; the feat of rescuing a man from a neighbourhood that he called 'home'. 'Since yesterday, we have moved three thousand Tamilians stuck in Kerala to their homes, and another two hundred Malayalis from Cumbum and its surrounding areas back to Kerala,' he said. 'Speak to them, boss, ask them if they faced any violence.'

The group immediately rushed to me and praised the TN policemen and TN people for all the kindness that they showered upon them. 'We had no troubles, we were staying safely in our Tamil neighbour's homes until police came to recue us. We love Tamil Nadu,' Ansari said, while making an effort to put up a smile.

An elderly woman member of the refugee group who had been staring at me for a while came closer to me and asked my name.

'Pradeep,' I said.

'Pradeep?' she wanted more. 'Are you a Malayali?'

She made me uncomfortable. I turned away and looked at my local source Kumar from Gudalur, another village near Cumbum, and spoke to him in Tamil to make the crowd surrounding me understand.

Roy and others, who had now been safely transported to where they belong (in Kerala), remained at the border as their home for the last twenty years had been in Cumbum.

*

On our way back to Cumbum, Kumar started talking about Roy and his business. 'His restaurant is very popular here as they serve authentic Kerala food. Although he knows most of us, his hotel was the first one to be ransacked three days ago when news about our women being molested and subjected to harassment reached Cumbum,' he said. 'Why did they attack Roy's restaurant despite his popularity in Cumbum?' I asked. He laughed. 'The first group that barged into the hotel and started ransacking the place was targeting him for personal reasons. They all had debts with him and he often abused them for not paying up. The men used this as an opportunity to warn him that he is an outsider here,' Kumar said. After pausing for a few minutes, he said that the later attacks—the hotel was attacked thrice over the last few days— were spontaneous and out of genuine concern for Mullaperiyar.

Kumar is 35, married, with two kids, and lives in a small hamlet outside Cumbum. Until that Monday, he owned a second-hand Tata Sumo that he purchased on loan from a local financer and ferried casual labourers from Cumbum, Gudalur, Pallarpatti and several other villages in Theni district to the cardamom plantations in Idukki district, Kerala for a price. Every morning at 7.00 a.m., more than a dozen women were crammed in his dusty Sumo to travel to work and they returned home later that evening. For ₹20 per trip, Kumar made just enough money to pay his dues and take care of the family. Between 9.30 a.m. and 5.00 p.m., he just lazed around in one of the many tea and wine shops in and around Kumily.

I came in contact with Kumar through an acquaintance in Coimbatore. He had set me up with a friend who then set me up with Kumar to help with my reporting. 'He is the best person to take you around, sir. He knows the place well and is also a victim,' they said.

Kumar's 2005-model Tata Sumo was one of the first few vehicles to be attacked in Kerala after Chief Minister Oomen Chandy decided to sit on a one-day fast to demand construction of a new dam. 'I got the first SMS that TN-registration vehicles were being attacked 3 kilometres away from Kumily inside the Kerala border at around 11.30 a.m. Immediately, I rushed to the estate and asked the workers I had brought in the morning to return to their villages as there might be problems. Our men were eagerly waiting to retaliate, they just wanted the 5,000-plus women labourers to return home before attacking all Kerala-registration vehicles and Keralites,' he said.

At around 1.00 p.m. Kumar left for Cumbum and was almost reaching the interstate border near Cumbummettu when a group of Malayali protestors blocked his vehicle. 'They were all armed with logs and hurled abuses at us. I asked the women to get out first and was still seated in the driver's seat when the first stone hit my windowpane and broke it. I ducked and asked the women to run into the fields. While the six or seven men who blocked my vehicle did not attack us, unidentified persons hurled stones and wooden logs at the vehicle from behind the bushes,' he said.

After all the women got out and escaped into the bushes nearby, the men clubbed the vehicle until all the glass windows were broken to pieces, seats were torn apart and the metal frame was severely dented. Before leaving, they punctured the tyres and hurled more abuses at Tamil 'pandis'.

'After guiding the women to the interstate border, I went to the Cumbummettu police station (in Kerala) to register a complaint, but they chased me away. I know those policemen well as I pass through this check post every day, but on that day they were unusually rude,' he said.

Kumar was seething with anger and I could empathize with him. While Kumar had nothing to do with the Mullaperiyar dam, his MUV, the only source of income for him, had been vandalized in the name of Mullaperiyar, and worse, the police were not willing to register a case against the perpetrators of this violence.

'So, what are you planning to do about it, your Sumo is damaged and they are not taking a complaint?' I asked.

'Don't know, sir, but I will do something. There are at least 500 vehicles with Kerala registration in Theni district, I am going to take one of the vehicles and start using it. They can have my Sumo, I don't need it any more,' he said.

I asked him if that would not amount to theft.

'In this part of the state, such events are common. My vehicle was ransacked and I was attacked. Yet, neither our police nor their police want to file a case. What else can I do?' he asked.

*

The violence increased manifold over the next few days. The lack of any kind of accountability only emboldened the criminals, who continued to vandalize anything that belonged to Keralites in TN and vice versa. According to the TN police, at least 500 vehicles including buses, vans and cars were damaged. Dozens of buses from TN and elsewhere that were transporting Sabarimala pilgrims were attacked by unidentified miscreants in Kerala, and every time a bus reached the TN border with damaged windowpanes and injured passengers, groups of villagers gathered around the victims and heard horrific tales of how they were attacked and abused verbally.

The villagers' anger at Malayalis was renewed with fresh vigour and they went about unleashing further rounds of violence, looting homes of Keralites and attacking their business establishments. During nights, groups of protestors searched for vehicles with KL registrations and burnt them down.

Since most Malayalis who lived in Theni district had either gone into hiding in their Tamil friends' homes or had left the state, the miscreants looted TVs, household utensils, motorcycles and everything else they could lay their hands on, and disappeared into the darkness. While the police acknowledged that there were random incidents of looting and arson, none of them bothered to register a case. 'We have not received any complaints so far, we will decide on registering cases whenever we get a complaint,' said a senior police officer.

Prakash from Annai Sathya Nagar in Cumbum is an estate owner. The 29-year-old bachelor owns two sprawling cardamom estates in Idukki district, Kerala, and employs over a hundred people, both Malayalis and Tamils. Although he is a Tamil, all his investments lie in Kerala but he has not been able to cross the interstate border to check out their condition since the violence broke out.

'My worst fear is that the Keralites might be looting my estate and home near Kumily under the supervision of the local police. Our men are doing just the same here, and when the Keralites get to know of it, they would rob all our properties,' he said.

'It is madness. These people seem to have lost their mind; I don't know what else to say,' he said, when I asked him how he felt about the violence.

Prakash had not stepped out of his home for the last two days, guarding his two new vehicles, a Mahindra Jeep and a Toyota Innova car, both KL-registration vehicles that he hid in his front yard in a cramped space in front of the living room. 'Although everyone here knows that I am a Tamil, I am not sure that my vehicles would still be safe,' he said.

The struggle that people like Prakash and Roy underwent for the last several years to ensure that there was a smooth relationship between Tamils and Malayalis for decades had been destroyed over a couple of days of violence. 'While it is important that we fight for Mullaperiyar water to continue to flow into TN for the benefit of our farmers, this madness unleashed by people here at the behest of some politicians has destroyed the trust that we have built over the years. This region will never remain the same. So much of distrust has crept in over the last few days alone that no Malayali would feel safe here anymore and no Tamil would dare to go to work in Kerala or to even run their businesses,' Prakash said.

*

Tired of having reported incident after incident in which angry Tamil protestors attacked and vandalized vehicles and properties belonging to Malayalis in Cumbum and Gudalur villages, which were the nerve centres of the protests, I decided to do some probing inland and went to Pallarpatti village, a few kilometres away from Cumbum where a memorial has been constructed for the architect of the Mullaperiyar dam.

When I reached the village, there was hardly a handful of men sitting under a concrete shelter that doubled up as the community hall during festivals and other events. As I went about asking questions to villagers about the importance of the dam and why they protested so fiercely against a new dam proposal made by the Keralites, a group of youth surrounded me.

'What's your name?' one of the boys asked.

'Pradeep,' I said.

'Full name sollu, sir.'

'Full name is Pradeep only, boss,' I said.

'What's your last name?' His tone now changed.

For probably the first time in my life, I was afraid of telling my surname. 'Pradeep Damodaran,' I said.

'Nee Malayalathan thane?' he asked.

Before I could answer his question, the boys came closer to me and said that they knew I was a Malayali. 'You better be careful and report properly, else you won't leave this place in shape,' one of them said. They took out a photograph of mine posted in Facebook and said, 'We know all about you and have already been warned.'

Shocked at seeing my photograph—and that too a bad one taken when I was drunk—yet not wanting to expose my fear, I remained at the same spot and tried to hold conversations with them. I explained to the villagers that I was born and raised in TN and that only my ancestors were Keralites. But they did not seem convinced.

Back in Cumbum, I was even afraid to get out by now. Rumour was that the locals wanted to attack anyone who even looked like a Malayali or whom they suspected to be one.

Despite being born and raised in TN, my name still remained Keralite, and that was apparently sufficient for the mischief mongers in Cumbum to attack me. Afraid of stepping out for fear of being attacked, I stayed indoors all day talking to sources over the phone.

For the first time in my life, I experienced how it felt to be afraid of one's identity. Despite being a Malayali born and raised in TN, neither Tamils nor Keralites were willing to trust me. On my way back to Cumbum, my source Kumar who had all along believed that I was a true-blue Tamil now suspected my motive. 'They are telling me that you are a Keralite and will be writing against us. Is that true, sir?' he asked.

I refuted this and claimed that someone was misleading him. But he was not willing to listen. 'Are you a Malayali?' he asked.

I denied that I was one and insisted that I be taken to my hotel room. Back in the hotel, my contact in Coimbatore called me on my mobile phone and requested me to rush out of the place. 'They are violent people. They are asking me if you are a Malayali. If they find out, then we will both be in trouble,' he said.

By now, I was growing sick and tired of explaining to people that I was, for all practical purposes, a Tamilian simply by virtue of having born in Chennai, and that too in one of the oldest hospitals of Chennai.

A little later, my friend from Coimbatore advised me not to be alone in the hotel or venture out. 'They are looking for trouble and might come after you. I suggest you leave the place immediately,' said Ashok, my friend.

I tried calling Kumar to tell him that I was leaving but he refused to pick up my calls. On the third night of my stay in Cumbum, I was forced to leave the place as I could not report freely there. The reason: I am a Malayali. It didn't matter to them that I was born in Chennai and that my parents were born in Chennai, and that my grandfather moved to Chennai even before India got independence.

*

On my way back from Cumbum to Coimbatore, I got a message that the KR Bakery, a popular bakery chain in Coimbatore owned by a Keralite, was attacked and that other Malayali establishments had closed shop fearing attacks.

Police said that the bakery chain was attacked by members of the group Naam Tamilar Iyakkam (NTI), who drove in motorcycles to the retail outlet of the bakery chain opposite Coimbatore Railway Junction, raised slogans against Kerala and hurled stones at the bakery and hotel. 'The entire incident lasted just a few minutes,' police said.

Three or four days after the escalation of violence in Theni district, several politicians, mostly belonging to small-time and Tamil chauvinistic parties such as Marumalarchi Dravida Munnetra Kazhagam (MDMK), Pattali Makkal Katchi (PMK), NTI and other outfits, thronged Theni district and made fiery speeches instigating people to fight against the injustice meted out to them. With the arrival of politicians, several other issues such as the Ealam war in Sri Lanka and the Koodankulam protests got connected with the Mullaperiyar dam row, and the issue transformed into a fight between Tamils and non-Tamils.

In Coimbatore, Chennai and rest of TN, several Malayali establishments were attacked in a similar fashion and the violence threatened to spread across the state. The entry of political and ideological outfits in the Mullaperiyar-related unrest gave the whole issue a new twist and threatened to blow into civil unrest.

Santosh (34) from Palakkad district runs a bakery in Vadavalli. He bakes good bread and biscuits. I am a regular at his shop and purchase most of my snacks, bread and biscuits from this bakery that is located close to my home.

Two days before Santosh was attacked, he had gone to his home town for a few days to attend a family wedding, closing down the bakery, when the attacks on Keralite establishments happened.

As he drove his KL-registration Tata Indica back from Palakkad to Vadavalli along the Coimbatore–Palakkad national highway, unidentified miscreants hurled stones at his vehicle in broad daylight near the Walayar check post, injuring him and his wife. His four-year-old son who was sitting in the back seat of the car was so traumatized by the attack that Santosh had to turn around and head back to Palakkad.

After reaching Palakkad and still unable to recover from the shock of the attack, Santosh called me to find out if it was safe to return. I advised him not to bring his car and instead come by train. For the next month or so, Santosh had to manage without a car as he was afraid to drive it in the city.

*

Palanisamy from Tirupur runs a small garment business and lives with his extended family. Every year, he takes his family out on vacation to different parts of the country, mainly to share the experience of exploring India with his two children.

Last year, he took his family on a tour to Kerala in the month of December during the winter holidays of the children. The ten-member family rented a Matador van from Tirupur for ten days and covered all the important tourist destinations in the state. All bookings had been done at least three months in advance. Palanisamy was a meticulous planner in business as well as in personal life.

During the last leg of his trip from Guruvayur to Tirupur, the van rented out by Palanisamy was attacked by unidentified miscreants. Fortunately, no one was injured.

I got to know about Palanisamy when he called my newspaper to seek help a couple of days after I returned from Cumbum. 'I am a regular reader of your paper and like the news you publish. I want you to print my story,' he said over the phone.

I wanted to hear his story.

'Sir, we were coming back home from Guruvayur to Tirupur via Palakkad when, somewhere along the way, our van was attacked with huge stones and wooden logs as we drove past a blind curve just a few minutes away from Guruvayur. Fortunately, the driver, who is a very experienced person, managed to get out of the scene without much trouble,' he said.

After recovering from the shock, Palanisamy, who clarified that he had no ill feelings for Keralites and even had several Malayali youths working for him, decided to file a complaint at the local police station.

'But when I went there with my family to file a complaint, they just refused to take it. I do not know Malayalam and they do not speak Tamil. But I think I managed to explain that my van was attacked and that we needed a case registered, at the least to claim insurance,' he said.

After waiting for over four hours in vain to get his complaint accepted, the tour party decided to return to Tirupur without registering the complaint.

'I have to answer to my travel agent and bear the expenses for repairing the van which could run into several thousand rupees. It is no fault of mine but I am helpless. Local police tell me that I have to register a case in the jurisdiction of the crime scene but the local police do not entertain us. What should I do?' he asked me.

I did not know how to answer that question. Several vehicles have been attacked, shops have been looted and homes have been ransacked. The police in both states have registered only a handful of cases to show that they were working and serious about the situation. But most complainants only shared Palanisamy's experience and I could do nothing about it.

'I am frustrated and I think I will join the local youth in my area and attack any KL-registration vehicle that comes to Tirupur. At least that would make me feel better,' he said.

We did not publish that story as I could not verify any of the details he had told us.

*

The next morning I read a report in the front page of an English daily about the Mullaperiyar anger spilling over onto the streets. While incidents of road rage between motorists and city buses during peak hours are a common affair in Chennai, this one had a twist.

The motorist and the bus driver who were engaged in an undeclared race along Anna Salai that morning eventually decided to vent out their anger at the traffic signal. The motorist drove his vehicle to the driver's seat of the bus and asked him to watch his driving in an impolite manner. The driver, upon noticing that the motorist was a Malayali, immediately hurled abuses not just at the motorist but all Keralites across the world and their treachery of not releasing the Mullaperiyar waters.

The motorist immediately retaliated by calling all Tamils pandi in full public view at the traffic junction. Soon, both got out of their vehicles to solve the Mullaperiyar dam row through a fist fight.

The passengers in the bus, who were in no mood for such a resolution, put an end to the madness and forced the driver to resume the journey.

*

2

Mullaperiyar Dam: A Legacy of Controversies

FOR TEN MONTHS in a year, somewhere deep inside the dense jungles of the Western Ghats in God's Own country, it rains. Moist clouds ram against the rugged slopes of tall mountains and burst, splitting into millions of giant droplets of water that thrash the earth, dampen its surface, and then flow gently down the slopes of the mountains.

When it is raining in Kerala, we know why people call this region God's Own Country. Life comes to a standstill here as water flows out of everywhere. Hundreds of thousands of tiny streams and rivulets flow down the slopes of the ghats and drain into the Arabian Sea, providing a livelihood for crores of people that live along their course.

Of these thousands of streams and rivers that flow in this region, only forty-four have a name. One of the longest among these rivers has been named the Periyar River (Periyar meaning 'Long River' in Tamil and Malayalam). Periyar originates from the Sivagiri Hills of the Western Ghats in Tirunelveli district of TN and flows westwards through steep valleys and gorges where several other tributaries join the river before it merges with the Arabian Sea at Munambham and flows into the Vembanad Lake at Varapuzha in Ernakulam district.

The river has a total catchment of 5,243 square kilometres of which just 113 square kilometres flows in TN. After traversing through dense forests for about 16 kilometres, the Periyar joins the Mullayar, one of its tributaries on its right, at an elevation of 850 metres. The river then turns west and cuts through the hills in a narrow deep gorge at about 11 kilometres below the confluence with Mullayar. This deep gorge is the site of one of the oldest masonry dams in the country which is also the one of the most controversial dams in modern Indian history. This dam, called the Mullaperiyar, is also probably the first interbasin river transfer ever attempted anywhere in the world, according to top structural engineers in the country.

The construction of this dam along the course of Periyar resulted in the formation of a reservoir spreading across an area of 8,000-odd acres right in the middle of one of the few tiger reserves in the country. The Periyar Reservoir and the Tiger Reserve are a popular tourist destination and one of the most important sources of income for people living in Idukki district, Kerala.

Water from this reservoir is diverted through a tunnel cutting through the hills to the TN border where it later flows through the Surliyar River into the basin of the Vaigai River in Madurai district, TN. The Periyar waters irrigate around 20,000 hectares of farmlands in Cumbum and Gudalur areas of Theni district in TN before joining the Vaigai.

After it reaches the Vaigai basin, these waters are diverted through a complex network of canals and tanks irrigating more than 1.8 lakh hectares of land along its way. Until the Mullaperiyar dam was constructed in 1895, this region depended solely on the much weaker northeast monsoon and was prone to frequent droughts. The water from the Periyar, that has been flowing into TN for more than 110 years now, is the lifeline for several lakhs of farmers who depend solely on these waters for irrigating their land.

The extraordinary vision of the proponents of the dam and the sheer feat of engineering that went behind its construction in successfully diverting a large west-flowing river to the east, cutting through a mountain range, which has been the claim to fame of the Mullaperiyar dam, has also been its curse.

*

Until November 2011, not many people in TN and Kerala even knew what the row over the Mullaperiyar dam was. While the farmers living in the five districts of TN, including Madurai, Theni, Dindigul, Sivaganga and Tirunelveli, and those living in Idukki, Ernakulam, Kottayam, Alapuzha and Thrissur districts of Kerala have been well-versed in their side of the controversy, there was no real clarity and holistic perspective over the issue in both the states, despite the Mullaperiyar dam row rocking the two state governments for decades now.

As one goes further away from the site of the dam, depending on which side of the Western Ghats we travel, the issue transforms and digresses into a plethora of allegations and counter allegations at the heart of which is a deep-rooted suspicion and resentment among the public over the neighbouring state and their own politicians who have let them down time and again.

The violent outburst of protests in Kerala during late 2011 and the massive riots that broke out in Theni district of TN that followed are a reflection of this deep-rooted feeling of bitterness against a system of governance which people still believe is not for their good and does not take care of their interests.

In December 2011, several political party workers and citizen groups in Kerala organized protests and even marched towards the dam to capture it from the hands of the TN Public Works Department (PWD) officials who are currently in charge of its maintenance. As a retaliatory effort, thousands of Tamil farmers threatened to march towards the dam to protect it from vandals as they feared that the Kerala police would allow its people to cause damage to the body of the dam. The protestors resorted to these measures as they had lost faith in their state governments and no longer believed that they would get justice from the courts.

To put it in a nutshell, on the one hand, the Mullaperiyar dam row is the story of an ageing dam constructed in the middle of a tiger reserve diverting a large river against its natural flow; and on the other, of thousands of poor people who have lost faith in their state government and in the legal system of the land.

So, what is this dam all about and why are these people so upset?

*

The Background

The Mullaperiyar dam was built by British engineer Colonel J. Pennycuick, commissioned during 1895. Since the dam was built on the land that belonged to the Maharaja of Travancore, a 999-year lease agreement was signed between the Madras Presidency and the Maharaja to lease out the dam site and 8,000-odd acres of land for the reservoir to the Presidency at the time of its construction in 1886.

The earthen dam that was built largely of a mix of lime surkhi is 1,200 feet long with the top of the dam standing 155 feet tall. It is capable of storing water up to a height of 155 feet, and a parapet of another 5 feet, that is, up to 160 feet exists at the top of the dam for almost 1,150 feet of its total length.

While a maximum of 155 feet of water could be stored in the Mullaperiyar, the full reservoir level, that is, the quantity of water that was allowed to be stored under normal circumstances was fixed at 152 feet by the engineers who designed the project.

The reservoir has a total storage capacity of 15.662 tmc out of which more than 10 tmc is dead storage. PWD presently manages and controls the operations of the dam, while security for the construction is provided by the Kerala forest and police departments.

To the left of the Mullaperiyar dam is the baby dam, a small dam constructed on a steep gorge nearby to hold the water from flowing downhill. The baby dam has a total length of 240 feet, and has also been built using the same technology as the Mullaperiyar dam.

To the right of the dam are thirteen spillways with floodgates that are capable of discharging floods at the rate of 2.12 lakh cusecs (cubic feet per second) in the event of flooding of the reservoir due to heavy rains.

For the last thirty-three years, the maximum storage level in the dam has been reduced to 136 feet after questions arose regarding its safety during 1979. Since then, the dam has stored only 136 feet of water, which is one of the chief flashpoints in the controversy.

*

For the last three decades, a large number of people living in Kerala have had a firm belief that the Mullaperiyar dam, which was built at a time when dam engineering was at its infancy and was constructed using inferior lime surkhi mix, is weak and could collapse any moment causing a disaster of catastrophic proportions. They fear that a collapse of the Mullaperiyar dam could release all the 15-odd tmc of water downstream to the Idukki dam sweeping everything on its way. They worry that the Idukki dam, which has a total storage capacity of 70 tmc, might not be able to hold this flood of 15 tmc of water and hence give way, which could wipe away a large chunk of the state including its top metro city, Kochi.

The government of Kerala and various political and non-political outfits have been pressing for the decommissioning of the Mullaperiyar dam and, instead, building a new dam using the latest technology to ensure the safety of the lives of millions who live in the state.

Since the issue was first taken to the notice of the Central Water Commission (CWC), the organization responsible for dams and other water bodies in the country, successive governments in TN have been repeatedly trying to prove that the dam is safe and have successfully convinced the CWC. At least three different committees have been appointed since 1979 by the CWC to inspect the safety aspects of the dam, and all three committees have so far concluded that the dam is safe. The maximum storage level in Mullaperiyar dam was reduced from 152 feet to 136 following fears among the public in Kerala during 1979. The agreement between the two states was that TN should carry out a series of strengthening measures to ensure the safety of the dam and then raise the water level back to 152 feet, so that its farmers get the full benefit of the dam as it was when designed. Despite carrying out a majority of the strengthening measures as prescribed by the CWC, neither the Kerala government nor its public have allowed TN PWD officials to raise the water level to 152 feet. The chief resentment among farmers in the Periyar–Vaigai belt of TN is that despite the TN government carrying out all the strengthening measures, they are being denied access to the Periyar waters that they are entitled to. The farmers and the TN government have been fighting for the last several years to raise the storage level at Mullaperiyar above 136 feet, and firmly believe that the dam is as good as a new one.

*

The months of April and May are the only time of the year when there is no rain in the Periyar catchment. Even if it does, it is just a mild drizzle and does not trickle into the 8,000 acres of land allotted for storage of water.

By the first week of June, the first symptoms of the arrival of the southwest monsoon reaches the cardamom hills in the west coast of the country with short spells of rainfall, and water begins to flow into the Periyar reservoir. On good years, the lake is brimming with water by the time the southwest monsoon shows signs of retreat three months later. By late August and September, there is a brief lull in the inflow but it picks up again during October as the northeast monsoon becomes active over the TN coast. The second season of inflow peaks by November, and on most years, the Mullaperiyar dam is almost filled to full capacity by this time, bringing cheer to lakhs of farmers and their families in the Periyar–Vaigai basin in Madurai, Theni and other districts in the region who would by now be assured of a good crop year ahead.

However, the news of Mullaperiyar dam brimming with water does not bring any cheer to the people living in Vallakadavu, Vandiperiyar and several other downstream villages situated along the banks of the Periyar River districts in Kerala. It triggers an uneasy feel deep inside their stomach, a fear that this might be the year when it may all came to an end.

For several decades, generations of Malayalis living in Idukki district have been living with this fear as every year draws to an end. Often, this fear transforms into frustration at their inability to do something about it and translates into anger. On some years, as it happened during 2011, this anger spills into the street in the form of violent attacks against Tamilians and abusive verbal discharge against Tamil politicians.

Suspicion and panic among the local populace, who live downstream the Periyar River in Idukki district, regarding the safety standards adopted in the dam has been around since the time the Mullaperiyar was commissioned and built. Ironically, the fear of an imminent dam disaster is almost as old as the dam.

Anyone who has visited the course of the Periyar River in Idukki district would not blame the locals' apprehensions. The deep gorges through which the river majestically flows, unleashing its fury especially during the monsoon season, could make any sane person wonder if a slim concrete structure would be able to hold the mighty river.

Although top structural engineers and hydrologists have repeatedly assured that the seepage of water through the walls of the dam was well within permissible limits and an indicator of the health of a good dam, the common public who live downstream find it perplexing that a significant amount of water is able to pass through several feet of lime surkhi and concrete and is able to reach the other side of the wall.

*

Some of the earliest press reports regarding the safety of Mullaperiyar dam appeared as early as 1925. Following these reports and rampant rumours among people living in Kerala, the Travancore government insisted that the safety standards of the dam be scrutinized. Upon the insistence of the Travancore government, the then Superintending Engineer, Trichy, C.T. Mullings rushed to the dam site to check the safety of the structure. He recorded that there was nothing wrong with the structure except for a few bank slips that could have taken place on the right bank surplus course and remedial measures were even taken up to stabilize the bank immediately.

In a press communiqué issue later by the government of Madras on 22 December 1925, the government clarified that reports appearing in the local media were misleading and that the authorities were taking necessary steps to improve and stabilize the surplus course to arrest further erosion.

The next wave of protests by anxious residents of Idukki district and the rest of Kerala took place in 1962 after a news item claimed that the Periyar dam was unsafe since it was old and had no scouring source to drain the reservoir fully.

Senior PWD engineers in TN attribute this anxiety to the sequence of inundations that were borne by the towns of Perumbavoor and Alwaye when the heavy floods of 1961 were discharged. Following these fears expressed by the media and downstream residents, the secretary to government, Trivandrum, sent a communication addressed to the secretary to the government of Madras suggesting a joint inspection of chief engineers of both states which would, according to them, go a long way in dispelling the fears of people living in Kerala.

Records indicate that the joint inspection was held on 10 April 1964 with the chief engineers, irrigation, Madras and Kerala, the superintending engineer, Kerala State Electricity Board and a director of the Central Water and Power Commission, New Delhi. A signed statement was also issued saying that the dam was quite safe and that the seepage was well within permissible limits. Several new safety measures were put in place, such as introducing wireless sets at the dam site and the executive engineer's office in Madurai and other electricity board and irrigation department offices in Kerala to communicate faster in the event of a disaster.

Fifteen years after those measures were taken, reports again appeared in a popular vernacular media in Kerala on 16 October 1979 that the dam was unsafe and there was serious structural damage due to erosion of lime over a period of time, and that it was time to think of a new dam at the site. This was mainly triggered off following the Morbi dam disaster that took place on 11 August 1979 when the Machhu-2 dam situated on the Macchu River in the state of Gujarat collapsed, and a huge wall of water swept away the town of Morbi killing anywhere between 1,500 (if the government figures are to be believed) to 20,000 (a figure the victims and dam activists claim) persons along its way.

Following the Morbi disaster, the leading vernacular daily pondered whether Mullaperiyar was the next dam disaster lying in wait and this created panic among the public. This led to an uproar in the state resulting in the issue being taken up in the Kerala Assembly, and the chief minister had to respond to the people's fear.

Although the chief engineer, irrigation, with the TN state government immediately clarified that the dam was safe and that continuous efforts were being made to maintain the dam in a safe condition, which included interactions with the Dam Safety Wing of the CWC, the issue attracted the attention of the Union government and the safety concern was forwarded to the chairman of the CWC in New Delhi.

The chairman of the CWC during that period, Dr K.C. Thomas, inspected the dam on 23 November 1979 along with the technical officers from TN and Kerala.

While reiterating to the public in Kerala that the dam was safe and that there was no need to panic, the chairman of the CWC suggested a series of strengthening measures to ensure that the safety levels in the Mullaperiyar dam was upto certain standards. The commission also required the TN PWD to reduce the storage level of the dam to 136 feet until all the advocated strengthening measures were completed.

The suggested strengthening measures were expected to be completed in three stages. Short-term strengthening activities that had to be done immediately included providing reinforced concrete capping after knocking off about one metre from the top of the dam, raising the shutters fully to limit the storage level of the dam at 136 feet, and provision of additional spillway capacity for controlling the water level in the event of a flood. This was to be followed by medium-term strengthening measures which included cable strengthening, and long-term strengthening measures which included strengthening the existing dam with a reinforced concrete backing on the rear face, provision of wireless facility to all concerned persons to ease communication in the event of a disaster, and several other measures were to be taken up to ensure that the dam was safe for a considerable amount of time in the future. One of the long-term measures that was suggested was for a joint team of engineers from TN and Kerala to explore the possibility of locating a new dam, within a month's time, as an alternative to long-term measures for strengthening the existing dam. The CWC was of the opinion that the water level could be raised to its original capacity of 152 feet after all the advocated strengthening measures were complete.

The TN PWD engineers took around twenty years to complete all the strengthening measures, which included cable anchoring and provision of reinforced concrete backing through the rear face of the dam, besides various other measures. While a majority of these measures were completed, the TN PWD officials could not complete the suggested strengthening of the baby dam, a smaller dam constructed to the left of the main dam, following opposition to the construction activities by the Kerala government. Since the recommendation of the CWC was that the water level in the dam could be raised to 152 feet after all the strengthening activities were completed, the TN government demanded that the water level be raised back to at least 142 feet as most of the strengthening activities were done as per the recommendations of the CWC. However, the Kerala government remained unconvinced about the safety of the dam and claimed that all the strengthening measures would be effective only when the water level in the dam was maintained at 136 feet, and not even a foot above this level. This has remained one of the core aspects of the conflict since then.

*

The matter went to the courts during the late 1990s. Separate writ petitions were filed at the Madras High Court and the Kerala High Court regarding the accepted level of storage at the Mullaperiyar dam. Due to the conflicting nature of the two cases and the resultant confusion that could occur if the two high courts gave conflicting judgments, the matter was referred to the Supreme Court of India (SC).

In the year 2000, the SC convened the meeting of the chief ministers of the two states to resolve the issue amicably. Since that meeting did not yield any positive results, the ministry of water resources constituted an expert committee to go into the details of the dam safety and revert with a report to the SC and advise the court on whether to raise the water level in the dam or not.

The expert committee consisted of CWC member B.K. Mittal as chairman, and included several eminent technical experts from various states including the Dam Safety Organization along with members from TN and Kerala. Based on the report submitted by the expert committee, the Supreme Court of India delivered its verdict in the case in 2006.

The judgment and its fallout are of great significance to the events that unfolded during the subsequent years, and in many ways are responsible for the conflict escalating to its peak during late 2011. Several important conclusions were reached in the SC verdict delivered on 27 February 2006 against the Writ Petition (civil) number 386 of 2001 between the petitioner Mullaperiyar Environmental Protection Forum and the respondents Union of India & Ors. The judgment was delivered by Chief Justice of India Y. K. Sabharwal and judges K. Thakker and P. K. Balasubramanyan.

In their petition, the petitioner had claimed that the 1886 lease deed was null and void and that it was entered in 'unholy' taste between the two parties. They also questioned the legality of the lease deed and the power of the SC in the issue, besides demanding that raising the water level to 142 feet would be bad for preserving the natural environment in the Periyar Tiger Reserve and for the safety of the dam.

One of the biggest grouses of those claiming for a new dam is that the 999-year lease deed signed between the British and Travancore government is null and void. The SC verdict of 2006 put that to rest, at least temporarily.

In the writ petition filed by the Mullaperiyar Environment Protection Forum, they had claimed that the agreements of 1886 and 1970 be declared null and void, and also that Section 108 of the States Reorganisation Act of 1956 be declared unconstitutional as it encroached upon the legislative domain of the State Legislature under Entry 17 of List II of the Seventh Schedule of the Constitution of India.

The SC, in response to that query, had stated that all interstate agreements would be continued by the parliamentary enactment, the States Reorganization Act. 'The state cannot claim to have legislative powers over such waters which are the subject of interstate agreement which is continued by a parliamentary enactment, namely, the States Reorganisation Act enacted under Articles 3 and 4 of the Constitution of India,' the verdict said.

'The effect of Section 108 is that the agreement between predecessor states relating to irrigation and power generation, etc., would continue. There is a statutory recognition of the contractual rights and liabilities of the new states which cannot be affected unilaterally by any of the party states either by legislation or executive action. The power of Parliament to make law under Articles 2 and 4 is plenary and traverse over all legislative subjects are as necessary for effectuating a proper reorganization of the states,' the verdict said, putting all claims about the validity of the 1886 lease deed to rest.

With regard to whether Article 363 of the Constitution bars this lease deed from the jurisdiction of the SC as it was made before the commencement of the constitution, the apex court has clarified that although the jurisdiction of the courts in respect of disputes arising out of any provision of a treaty, agreement, covenant, engagement and other similar instruments entered or executed before the commencement of the constitution was barred in the manner provided in Article 363 of the Constitution of India, the main reason for ouster of jurisdiction of courts was to make certain class of agreements non-justiciable and to prevent Indian rulers from resiling from such agreements as that would have affected the integrity of India. 'This Article has no applicability to ordinary agreements such as lease agreements, agreements for use of land and water, construction works. These are wholly non-political in nature. The present dispute is not in respect of a right accruing or a liability or obligation arising under any provision of Constitution,' it said.

While addressing the core issue of whether raising the water level of the reservoir from 136 feet to 142 feet would result in jeopardizing the safety of the people and also degrade the environment, the court gave the verdict that the water level could be raised from 136 feet to 142 feet, and that raising the water level to such an extent would not affect the flora and fauna of the Periyar Tiger Reserve and that it would only benefit them. The court also mentioned that the Kerala government had adopted an 'obstructionist approach'.

*

The writ petition was disposed off after permitting TN to raise the level of water to 142 feet and continue with the suggested strengthening measures. The verdict, as obvious, was a big blow for the Kerala government and the Mullaperiyar Environment Protection Forum, and this was reflected in the mood of the public living in Idukki district who felt let down by the SC and their own engineers and politicians who had been claiming for decades that the Mullaperiyar dam was unsafe and could collapse any moment. Several hunger strikes and fasts were organized and the Kerala government took some aggressive measures to ensure that the water level was not raised beyond 136 feet.

*

The SC verdict brought cheer to lakhs of farmers in the Periyar–Vaigai basin of TN and celebrations broke out in the region as after twenty-seven long years, bountiful amounts of water would again flow into their farmlands. But their hope and joy lasted for barely three weeks.

In TN, it was time for yet another assembly election. Chief Minister Jayalalithaa was in no mood to the celebrate the court victory and raise the level of water in Mullaperiyar dam to 142 feet until she got elected for another term, although the farmers' associations were eagerly waiting for the waters. Hence, the water storage level at the dam remained at 136 feet despite the SC order for the next few weeks.

On 18 March 2006, the Kerala State Assembly passed a controversial amendment bill to the Kerala Irrigation and Water Conservation Act, 2003, which was in direct conflict with the SC verdict in the Mullaperiyar issue.

As public resentment against the SC verdict gained momentum and poured out onto the streets, the Kerala government had to do something to convince the people that they were concerned about their citizens. The controversial amendment was intended to achieve that objective.

In the amendment bill, two new sections, Section 62A and 62B were introduced by the Kerala Irrigation and Water Conservation Act, which fixed a ceiling for the full reservoir level (FRL) of twenty-two dams in Kerala that were listed as endangered due to their age, degeneration, degradation besides having structural and other impediments. The Mullaperiyar dam was the first one in the list of endangered dams and its FRL was fixed at 136 feet.

The amendment also introduced another section 68A that provided the Act and its authority protection of action and immunities from challenge. Since this new section ensured that the SC order would not be effective in determining the FRL of endangered dams as per the Act, it is reproduced verbatim.

Section 68A Protection of action and immunities from challenge, etc. : *(1) No suit, prosecution or other legal proceedings shall lie against the Authority or against any officer or employee in respect of anything which is done in good faith intended to be done in discharge of official duties under this Act.*

(2) No civil court shall have jurisdiction to settle, decide or deal with any question of fact or to determine any matter which is by or under this Act required to be settled, decided or dealt with or to be determined by the Authority under this Act.

As news spread among the farmers in TN over the enactment of a bill that denied them the right to 142 feet of Mullaperiyar waters despite a SC order, their resentment once again spilled out to the streets and resulted in traffic jams and road blocks that lasted for a few days. Emotions ran high among the farmers in Madurai and Theni district.

Following the amendment of the Kerala Irrigation and Water Conservation Act, the TN government once again approached the Supreme Court of India seeking its intervention in implementing the court order of 2006 allowing the Mullaperiyar dam storage level to be increased to 142 feet, besides questioning the legality of the amendment which was in direct conflict with the SC order.

The constitutional bench of the SC issued an interim order in September 2006 in which the two state governments were asked to hold discussions on the issue and try to solve it amicably with or without the intervention of the Union government. Following this order, the chief ministers of the two states met and discussed the issue. Despite several such meetings, no consensus could be arrived at.

*

During the intervening period, the Kerala government had assigned independent experts from IIT Delhi and IIT Roorkee to study the hydrological and seismological safety of the dam as the state government felt that the CWC's opinion on this issue could not be relied upon.

The civil engineering department of IIT Delhi conducted a detailed study on the probable maximum flood (PMF) estimation and flood routing study for the Mullaperiyar dam and submitted its report to the Kerala government. The study concluded that the dam was hydrologically unsafe for passing the estimated maximum probable flood with the existing spillway capacity.

The seismological study on the region in which the dam is located and on the structural stability of the Mullaperiyar dam considering the seismic effects was conducted by the Department of Earthquake Science in IIT Roorkee during May 2008.

Some of the conclusions of the study are listed below: The

Mullaperiyar dam site lies on the western coast of India in the State of Kerala. It lies in Seismic Zone III as per the seismic zoning map of India where a maximum intensity of VII is expected. As the Mullaperiyar dam is more than 110 years old, constructed in stone masonry in lime surkhi mortar, it was envisaged that this old dam will be vulnerable under a future strong motion earthquake in the region and in the eventuality of dam failure, it may result in human and economic losses.

*

Armed with these two studies that turned out in favour of their argument that the Mullaperiyar dam was unsafe and that a new dam had to be built, the Kerala government went ahead with its proposal for the construction of a new dam and efforts were taken to locate a suitable site for the same.

On 14 August 2007, the Kerala cabinet approved a proposal to start preliminary work on a new dam at Mullaperiyar. As news spread regarding this cabinet approval, violence broke out once again in the TN–Kerala border as the members of the Periyar–Vaigai Farmers' Association demanded the TN government block all trade with Kerala until it gave up the proposal to build a new dam in the Periyar basin.

However, the Union Ministry of Environment and Forests granted permission to the Kerala government to conduct a survey and identify a suitable site for a new dam inside the Periyar Tiger Reserve during September 2009. The TN government immediately approached the centre, and later the SC, against this approval given by the Ministry of Environment and Forests. However, their plea was rejected.

In February 2010, the Supreme Court of India constituted another empowered committee, consisting of representatives from TN and Kerala besides eminent engineers from the CWC and other central organizations, to study the safety issues of the dam once again in an effort to allay the fear in the minds of the public in Kerala. The court would also study the legality of the Kerala government's legislation overriding the SC order in fixing the FRL for Mullaperiyar dam at 136 feet. Thus, the Mullaperiyar dam row went right back to where it all started ten years ago.

*

In November 2011, the Mullaperiyar region received unusually heavy rains and the dam was filling fast. As with every other year, fear over the stability of the dam spilled over onto the streets. The public outrage was complemented by mild tremors that shook Idukki district on 18 and 26 November and gave way to an outpouring of protests demanding the decommissioning of Mullaperiyar dam once again in Kerala.

Besides these factors, the screening of a film *Dam 999* with the story revolving around a dam disaster in Kerala all led to spontaneous outbursts of protests and a revival of the shrill pitched cry for the construction of a new dam.

On 3 December 2011, members of the Youth Congress in Kerala staged a protest march to the Mullaperiyar dam and the TN Electricity Board office located next to it. The next day, hundreds of BJP workers marched to the dam with an intent to break it. While these were mere political gimmicks played out by party workers at the insistence of their leaders, there were also several attacks on Tamils and vehicles bearing TN registrations in and around Kerala. As it was the peak season for the famous pilgrim site Sabarimala where thousands of Tamils visit every winter, hundreds of TN-registration vehicles were stuck in Kerala, many of which were attacked at random by groups of miscreants.

Buses plying between Kerala and TN carried derogatory messages against Tamilians, and the TN chief minister and police on both sides of the border were actively pressed in defusing the tension.

When the Kerala chief minister decided to sit on a one-day hunger strike demanding the decommissioning of Mullaperiyar dam, the Youth Congress and other activists in Kerala blocked all roads to TN and over 500 Tamil estate workers, who travelled from Cumbum and Gudalur in Theni district to Kumily, were struck on the border. As rumours spread about the molestation of women, violent riots broke out in Cumbum and Gudalur where homes and business establishments of several Malayalis were ransacked and looted.

Even as violence escalated in both states, the Kerala assembly appealed to the prime minister to reduce the level of water in Mullaperiyar dam to 120 feet. The protests that were somewhat manageable until then broke into a full-fledged riot as thousands of farmers and others living in Theni district marched towards the Kerala border with sticks, logs and knives claiming to protect the dam against a possible attack from Keralites. The march attracted thousands of volunteers, and as hate messages spread, vehicle movement and trade between the two states came to a grinding halt. The protests and rioting lasted for a month.

The situation returned to normal during early January 2012 after TN farmers were assured that the water level in the Mullaperiyar dam would not be reduced below 136 feet, which has been the storage level since 1979.

*

By early 2012, the Empowered Committee (EC) of the SC had completed its technical study and had once again reiterated that the dam was hydrologically, structurally and seismologically safe for the coming future in May 2012. This report released by the EC once again angered the residents of Idukki district and Keralites in general.

Following the results of the EC report, the Kerala state government once again announced that it would go ahead with the proposal for a new dam, as had been announced in the annual state budget during 2011, with the allocation of a sum of ₹500 crore.

*

Critical Issues in the Conflict That Have Never Been Discussed in a Public Forum

While there are several dams in India that are as old as the Mullaperiyar, none of them have received this kind of attention and scrutiny.

At the heart of the Keralites' discontent over the safety of the Mullaperiyar dam is the awareness that they are risking their lives with a 116-year-old structure out of which they receive little or no benefit at all. While the beneficiaries of the Periyar project live hundreds of kilometres away in TN, the victims of a dam disaster would be the citizens of Kerala who stand to gain nothing from it.

As a senior Kerala-based structural engineer put it succinctly, one always has to do a cost-benefit analysis, and in the case of the Mullaperiyar dam, the Keralites feel that the equation seems quite out of balance.

It's All about Money

The Periyar project is governed by the Periyar Lease Deed of 1886 signed between the Maharaja of Travancore and the Secretary of State for India in Council, leasing the dam site and around 8,000 acres of catchment area and the water stored in it to the British government for a rent of ₹40,000 for a period of 999 years.

The lease was later modified in 1970 to include the Periyar Hydro-Electric Project in the deed, and the State of TN agreed to pay Kerala ₹12 for every Kilowatt year of electricity generated using the Periyar waters. The rent for the land used by TN was also increased to ₹30 per acre of the leased land.

According to the Kerala government, the state generates revenue of only ₹8.99 lakh every year, which includes the sum of ₹2.58 lakh that it receives as rent for 8,592.97 acres of land leased out to TN and ₹6.41 lakh that it receives as royalty from the TN government for the power generation.

The popular feeling in Kerala is that while TN generates crores of rupees using the Mullaperiyar waters through irrigation and power generation, Kerala receives a pittance and has to stand the risk of a century-old dam. The general perception in Kerala is that the price they receive for holding on to an ageing dam is grossly inadequate. At the heart of the Mullaperiyar controversy is this financial dispute.

Power Situation in Kerala

While the power situation in Kerala is not as bad as it is in TN, which was reeling under a severe power shortage of 3,000 MW as in May 2012, and even in 2014, there is a looming crisis in the neighbouring state as it does not have any major projects lined up for the near future.

According to the statistics disclosed by the Kerala State Electricity Board (KSEB), the state presently has a demand of around 3,500 MW of electricity, which is expected to soar up to 4,534 MW by 2016–17, as per the 17th electrical power supply forecast conducted by KSEB.

Against this rising demand, the KSEB presently generates only a total of 2,234.4 MW of electricity, out of which 1,997 MW is generated through hydroelectric power, that is, almost 90 per cent of the generation capacity. The state also receives around 633 MW of electricity from central projects, putting the total installed capacity at around 2,800 MW.

With no new major projects in the pipeline and the gap between demand and supply almost poised at around 1,000 MW, the state is purchasing a huge quantity of electricity from other power surplus states and through private vendors at a cost of approximately ₹8–9/unit. The price, however, varies through the year and can rise up to even ₹15/unit depending on the demand.

While the KSEB is purchasing power from outside vendors to fulfil its needs at a huge cost every day, it is only natural for the state government to turn their attention towards the Periyar Hydroelectric Power Project, where the Periyar waters are used to generate around 140 MW of electricity for which Kerala state receives merely ₹0.12/unit as per the 1970 modified lease document.

The Kerala government has the responsibility of fulfilling the state's power needs and has been eyeing the Periyar waters to generate an additional 150-odd MW of electricity for its own use.

Meanwhile, TN had a power demand of 11,000 MW every day, while the installed production capacity as in 2012 was only around 8,000 MW. Although there was a 3,000 MW difference in demand and supply, the crisis was somewhat abated in 2014, with measures taken by the TN government to generate additional electricity through the various thermal and other power projects that are under production.

With the energy availability in TN is almost four times that of Kerala, the state could consider doing away with 140 MW of power generated at the Periyar Hydel Plant and thus solve a major issue behind the whole controversy.

TN Farmers' Dependence on the Mullaperiyar Water for Over a Hundred Years

Another major impediment to arriving at a solution to this issue is the continued dependence on lakhs and lakhs of TN farming families on the Mullaperiyar waters. A hundred and twenty years ago, the British government decided to divert the Periyar waters to Theni, Madurai, Sivaganga and other dry districts in TN to help people in this part of the land find a livelihood.

Without a doubt, that could have been the smartest thing to do to find a steady way of living for a warring and violent community without a steady occupation. It was a time when there was no industry to speak of, no technology available to transfer water through long distances without seepage, agricultural research was at its infancy, and the country was not independent.

The past hundred years have enough examples of successful cities that have developed with little or no water source. The vast advances in technology have drastically reduced man's dependence on agriculture as the key to civilization. Technology parks, industrial estates and commercial centres have all compensated for the lack of agricultural land in providing a decent and legitimate means of livelihood even in various parts of TN.

Strangely, the farmers and a majority of rural residents of Theni, Madurai and other districts benefitted by the Periyar waters still depend on these waters as their only source of livelihood and continue to have a strong bond with the Periyar waters. Successive governments in TN have done nothing to promote industry or commerce in these regions as has been done in Chengalpattu, Kancheepuram and other areas where they could have still continued agriculture as there is reasonable water source in these areas. Lack of a long-term vision and an absolute dependence of TN farmers on Mullaperiyar waters have compounded the problem and increased the emotional quotient of the conflict.

*

The two states have not even recognized each other's problems and see only their individual points of view. While the Kerala government has been fighting for decommissioning of the dam and construction of a new one, they have never bothered to discuss the controversy over the maximum storage level, which is currently fixed at 136 feet despite a court verdict on raising it to 142 feet.

TN, on the other hand, has never even agreed to the fact that the Mullaperiyar dam might have to be decommissioned and rebuilt at some point in the future as it cannot stand good for a thousand years. As far as the TN government is concerned, the entire controversy is over increasing the storage level in the dam to 142 feet, as had been approved by the SC in 2006 and 2014, and later raising it to 152 feet, which is the designed maximum storage level.

*

3

One Thousand Years of Lease

BESIDES THE DISAGREEMENT between TN and Kerala over the safety of the Mullaperiyar dam, the next most controversial factor in the tussle between the two states is the 999-year lease deed signed between the Maharaja of Travancore and the Secretary of State for India in Council during 1886.

According to the original lease, the site of the dam and 8,000-odd acres surrounding it, that is, the land surrounding the Periyar River bounded by a contour line 155 feet from the deepest point of the river at the dam site, was leased to the British government for a period of 999 years for a rent of ₹40,000 (at the rate of ₹5/acre of land leased).

The lease was, however, later amended in 1970, after a series of discussions between the two state governments, when the lease rent was raised from ₹5/acre to ₹30/acre of land used by TN which came up to ₹2.4 lakh approximately. Besides, the hydroelectric project and related payment details were added as a supplement to the original deed.

Even if one were to set aside the resentment in Kerala over the lease amount that had been negotiated in 1886 and which is largely seen as highly unfair, the lease term that extends up to 999 years, with an additional provision to renew it for another 999 years from the time of expiry of the first lease term if the British government wished to do so, only makes the lease deed even more absurd.

It was about a thousand years ago that the land now called as India and Pakistan was first attacked by Mahmud of Ghazni. After Ghazni and Ghori came so many invasions, then the Mughal Empire and much later the British. To expect governments and people to abide by a lease for such a long period of time in history simply defies common sense.

As bizarre as it is, 999-year leases have been quite common in Britain and in other Commonwealth countries where the British system of governance was followed. Some of the prominent properties that have been leased out for a period of 999 years include the Queen's Park in Toronto and Ontario, Canada, that was leased by the University of Toronto to the Province of Canada between 1859 and 2858 and the Hampshire County Cricket Club land located in Hampshire, England, that was leased by the Queen's College to the Hampshire County Cricket Club for a period of 999 years between 1996 and 2995.

One of the main reasons for Kerala's resentment regarding the lease is due to the fact that the original lease was signed during a different era between the powerful British government and a feeble princely state. The British Empire's well-planned exploitation of its colonies and the adverse impact that this colonization has had on these countries including India is well documented and common knowledge now. Hence, an agreement signed between the British Empire and a small princely state down South, that depended on the empire for trade and other activities, could not have been a fairly negotiated one.

Records indicate, though, that there was a significant amount of negotiation done by the British and the Travancore government for over a period of twenty years before the actual lease document was signed.

*

Some of the early discussions regarding leasing out land from the state of Travancore were taken up as early as the 1860s when Captain Ryve conducted his field study and submitted his report. In 1862, the then Diwan of Travancore Raja Sir T. Madhava Rao wrote to the resident in Travancore saying that the project appeared to be fraught with promise; and after closer examination and preparation of plans and estimates and if the work was found to be practicable and profitable, there would be no difficulty in arranging fairly the interests of the Travancore and Madras governments with regard to the outlay of profits.

There had also been discussions between the two governments to build the dam as a joint venture project and share the profits between them as is evident in the note written by the then secretary to the government of Madras to the Travancore government stating that it was prepared to assign to the Travancore government one half of the net of profits which could arise from use of the waters.

However, later, the Travancore government decided to pull out of the joint venture, considering the expensive and riskful nature of the work, and agreed to accept a sum of ₹75,000 a year for the site of the reservoir and use of water stored in it in a letter sent to the British government dated 24 October 1873.

The negotiations over the lease amount went on for another decade or so during which period there was also a suggestion to transfer 8,000 acres from British territory to the Travancore government in exchange for the land leased out to them. However, these proposals were subsequently rejected by the Secretary of State for India.

After years of negotiations, the terms of lease were finalized in 1885 and the lease was signed in 1886. Some of the important terms and conditions laid down by the Travancore government for signing the lease included that all lands that would be submerged (approximately 8,000 acres), and any more land that could be used for construction and preservation of the dam, would be granted to the British government on a perpetual lease for an annual rent of ₹40,000 that was payable to the Travancore government for use of the land and the water. The payment was expected to commence from the year when the Periyar waters would be diverted to British territory.

The Travancore government also insisted that if the work was abandoned or remained unused for some reason, or if water was not diverted, the land and the water available on the land would revert back to the Travancore government, while the British would not be obliged to pay any further rent. It was also made clear that the Travancore government would continue to exercise all its rights of sovereignty over the lands leased and its inhabitants.

While the British were handed over the rights to do fishing on the reservoir and all the ponds associated with it, besides being allowed to fell timber and bamboo required for the construction and preservation of the dam on a permit basis, the Travancore government retained all rights on the mines and minerals that may be discovered in the land.

After the lease deed was signed, the Madras government conducted a detailed survey of the land and demarcated lands to be counted against the lease, both in the reservoir and the camps. The Travancore government raised queries and insisted that their representatives be present while surveying. A joint inspection of the areas was conducted and the total extent of land that fell under the lease deed was clearly determined and demarcated on the ground.

*

The first dispute over the lease deed of 1886 arose during the early 1900s when the Government of Madras wanted to generate hydroelectric power from the Periyar waters that flowed through the tunnel and fell into the Vairavanar Stream of the Vaigai River. The two governments disagreed over whether the lease deed of 1886, made in connection with the Periyar Irrigation Project, had conferred the right to generate electricity from the waters flowing through the mountains to the British government.

The matter went before the Periyar Arbitration Tribunal consisting of Sir M. David Devadoss, an ex-judge of the Madras High Court and Diwan Bahadur V.S. Subramania Aiyar, ex-Diwan of Travancore State. While Sir C.P. Ramaswamy Aiyar was the counsel for the Travancore government, Sir Alladi Krishnaswamy Aiyar was the advocate general representing the Madras government.

The case of the Travancore government was that the waters of the Periyar reservoir were, according to the terms of the lease deed, to be used only for irrigating lands and for no other purpose, and hence, the use of water for generation of electricity could not be done without a fresh agreement.

The Madras government pleaded that the waters of the Periyar lake were the absolute property of their government and that they may put them to any kind of use after they entered British territory, and as all the electrical works would be located on the eastern side of the British territory, the Travancore government was not entitled to raise any objection to the execution of the proposed hydroelectric project.

After hearing the counsels, differing awards were passed by the arbitrators on 16 January 1937. Sir David Devadoss declared that the execution of the proposed hydroelectric project would not violate the terms of the existing deed of 999 years, and that the Travancore government could have no legitimate ground of complaint with reference to power generation.

To substantiate his point, Sir David Devadoss noted that even as far back as 1862, the idea of utilizing the Periyar waters for generation of hydroelectric power after they were brought to the Madras side of the ghats was present in the minds of the officers of the British government. Since there was already a discussion on power generation, the arbitrator stated that the Travancore government should have made a claim for it even at the time of negotiating the lease back in 1886, and that it was done as the Travancore government knew that it could not sustain its claim.

While indicating that no principles of international law or conventions were applicable to the case of a grant by one state to another, Sir David also pointed out that the entire row over power generation rights in the contract was for financial gain. To put it in his words, 'Notwithstanding the disclaimer one cannot resist the inference that the real object of the objection to the hydroelectric scheme is a desire to get something more from the Madras government than the rent reserved in the agreement or to get a share of the profits that might accrue from the scheme.'

Another important point that he put forth was that when the land was leased to a person for forming a reservoir for storing water and drawing from it as he or she needs, it could not be said that the grant was anything less than the full property in water.

The other arbitrator, V. S. Subramania Aiyar, however, differed and held that since the nature of the grant was a lease, it was limited to the terms and conditions in the lease. 'Hence it was a case of the grantee taking such rights as are given and the residium of the rights continued to reside with the original proprietor. He may restrict the purpose for which the property may be used or the manner in which it should be used,' he noted. 'If the use was restricted to a purpose, diversion to some other purpose, even if it does not entail the use of a greater quantity of water, will not be allowed.'

Highlighting that despite the period of lease being 999 years considered as a perpetual term, it was still a lease despite the longevity of the lease term. 'The grant of a right to use the water for one of the natural purposes cannot be taken up to imply the grant of a right to use it for an artificial or commercial purpose,' Mr Aiyar pointed out. 'For the generation of electrical power, the water had to be available at a high level and fall and the owner of that water in that height in the bed of the Periyar with that potential was the Travancore government.'

Subramania Aiyar concluded that neither under the deed nor by virtue of any right acquired under the deed was the Government of Madras entitled to use the water of the Periyar Project for generating hydroelectric power. They have acquired the right only to use the water for irrigation.

As the arbitrators differed in their opinion, the matter was referred to the umpire, Sir Nalini Ranjan Chatterjee, before whom arguments were made during January 1941 by Sir C. P. Ramaswami Aiyar and Sir Alladi Krishnaswami Aiyar.

The umpire's findings declared on 12 May 1941 were: (i) upon a construction of the lease document, the lessee has the right to use the water for irrigation purposes only; (ii) that the lessee has no right to use the water for any purpose other than irrigation; and (iii) that supposing it is possible to use hydroelectric energy for carrying or distributing water or doing any other act in connection with irrigation, the lessee has the right to generate and use hydroelectric energy for irrigation purposes only.

Thus, the Madras government lost their case.

*

Despite the setback in the arbitration, the governments of Travancore and Madras started fresh negotiations to arrive at an agreeable set of terms and conditions to start the hydroelectric project.

During the negotiations, the Government of Madras relinquished the fishing rights granted in Clause VI of the Periyar Lease Deed in favour of the Government of Travancore. The Travancore government also put forth a proposal to create a National Park in the Periyar catchment above the reservoir, which was also accepted and a wildlife sanctuary came into being. The right to run a tourism industry by the Government of Travancore utilizing the Periyar lake precincts was also granted along with the right to ply their boats without obtaining the permission of the Government of Madras since then.

The Government of Travancore insisted on an amendment to the main lease deed in respect of the rent for the land that had be handed over to the Madras government, and the rent was revised from ₹5/acre to ₹30/acre. The Travancore government also insisted that the rate be revised every thirty years subject to the prevailing conditions.

Having successfully negotiated all these amendments to the original deed, a new amendment and a supplementary deed for the hydroelectric power generation were signed in May 1970. As per the agreement, the Government of Kerala handed over the full right, power and liberty to construct head works, tunnels, pumping installations, waterways, transmission, distribution, telephone lines, etc. to the Government of TN for generation of hydroelectric power at the Periyar Power House.

The Government of Kerala was entitled to receive a payment as follows:

(i) When the electrical energy generated by the Government of TN at the Periyar Power House did not exceed 350 million units in a year, a sum of ₹12 per kilowatt year of electrical energy was to be received.

(ii) When the electrical energy generated at the power house exceeded 350 million units in a year, the Government of Kerala would receive ₹12 kilowatt year for the first 350 million units, and a rate of ₹18/kilowatt year for the electrical energy generated in excess of 350 million units.

(One kilowatt year meant 8,760 units of electricity for the purposes of the contract)

Besides the portion relating to generation of hydroelectric power, the clauses granting fishing and tourism-related rights to the Government of TN in the 1886 lease deed were modified, and the right to fishing and promoting tourism-related activities in the reservoir were granted to the state of Kerala. The annual rent was also altered to ₹30/acre of leased land from ₹5/acre.

While this amendment to the original lease deed was not considered a big deal in Kerala at the time, the present day politicians and people believe that the then state government had completely hoodwinked the people of the state by signing the 1970 revised lease deed without debating it in a public forum. They claim that the Kerala government was pressurized to agree to a deal that was highly unfair.

*

Again, in 2002, a writ petition was filed by the Mullaperiyar Environmental Protection Forum, a citizens' group from Idukki district, against raising the water level in the Mullaperiyar dam to 142 feet. In the petition, they had also claimed that the agreements of 1886 and 1970 be declared null and void and that the lease deed be declared unconstitutional as it encroached upon the legislative domain of the State Legislature. The petitioner had raised objection about the legality of the agreement between the Maharaja of Travancore and the British government. It was claimed that the agreement was entered into in 'unholy' haste and was virtually thrust upon the Maharaja who was forced to accept it.

However, the SC verdict of 2006 that was delivered in favour of TN farmers clarified that the 1886 lease deed was, in fact, valid and could not be declared null and void. The judges had also clarified that since the agreement between the British government and Indian rulers was wholly non-political in nature, it would continue to remain operative.

Despite the issue being studied in detail by the panel of judges in the case, there is still resentment among the legal fraternity and the political top brass in Kerala that the lease deed should be examined in more detail by a larger constitutional bench to confirm its legality, and they were pressing for the same when the case comes to trial in 2012.

*

4

How Safe Is the Mullaperiyar?

UNTIL A FEW years ago, my knowledge on dams was limited to lessons learnt during high school geography classes: dams are giant walls built across rivers to store water for irrigation and power generation. Period. As for safety of dams, if these walls are weak, then it would break.

For a chemistry graduate-turned-software professional, this limited understanding of dams was sufficient to move on with my life. To most people who are not structural engineers or have studied dam engineering in detail, including the people living downstream Mullaperiyar and elsewhere in Kerala, dams are still giant walls that would break if they are weak.

Over the last few months of my interactions with dam experts and top civil engineers in the southern part of the country, I have been humbled to understand that there is a lot more to dam engineering than just building strong walls across rivers. Let me make an attempt to retell the story of safe dams.

In the context of gravity dams such as the Mullaperiyar, safety aspects that are of primary concern include hydrological, structural and seismic safety. Statistics indicate that a large number of dam failures across the world occur due to hydrological failure and, hence, it is considered the most important factor while designing dams.

According to structural engineers, a dam is considered hydrologically safe when it is able to discharge flood waters effectively without it having to flow over the top of the dam, when the possibility of a dam break is extremely high. To discharge flood waters effectively, dams are provided with spillways or flood gates that divert the water flowing into the dam out of the catchment, maintaining the water levels within safe limits. The capacity of the spillway is designed to accommodate the maximum possible flood in a particular region, that is arrived at using historical data of rainfall in the region besides other mathematical calculations.

Structural safety of a dam is to ensure that the construction is made of strong material as per accepted standards. The geography of the location and the condition of rock at the foundation of the dam is extremely crucial in determining the structural safety of dams. As for seismic safety, it simply means that the dam should be able to withstand earthquakes that are likely to occur at the site of the dam.

A hundred and twenty years ago when the Mullaperiyar dam was built, there were no real guidelines to ensure safety of dams. Over the years, several organizations, both national and international, have laid down strict guidelines to be adhered to for dam builders to ensure that their dam does not become a 'water bomb'.

Globally, the International Congress on Large Dams (ICOLD) has been the pioneer organization in analysing various aspects of dam engineering, since its inception, to ensure proper design and construction of safe dams. ICOLD holds several interactions of dam experts from across the world to share and update their knowledge on the latest in dam technology to ensure that hazards are reduced to a minimum.

In India, the CWC operating under the ministry of water resources had formed a Dam Safety Organization in the year 1979 to assist state governments in various activities connected with the safety of dams in their limits.

Soon after it was formed, the Dam Safety Organization had initiated action for reviewing the existing procedures of dam safety in the country and also to evolve appropriate safety practices. The report produced by the Dam Safety Organization, by and large, determines the guidelines for determining the safety of existing dams. Each state government has also laid down individual guidelines concerning the safety of dams in their respective areas as per the broad guidelines provided by the CWC.

In 1982, the Government of India, Ministry of Irrigation constituted a standing committee to review the existing practices and to evolve unified procedures of dam safety for all dams in India, under the leadership of the Chairman, CWC. The report submitted by the Committee on Dam Safety Procedures continues to be one of the guiding documents whenever the question of stability of a dam arises although, as mentioned in the report, the standards set by the ISI and ICOLD are also considered as and when found appropriate.

As part of the process of framing guidelines for various safety aspects of dams, the CWC, for the first time, formulated a design flood criteria for hydraulic structures in the early 1960s. This criterion was circulated to various states and was finalized in 1968.

With regard to major projects, the CWC utilized these design criteria for ensuring, within workable limits, that its guidelines are followed in all new projects. However, in minor and medium projects, which were either not examined or not examined in detail, conformity with the criteria would not be adequately ensured. Also, the various existing projects and projects under construction did not necessarily conform to the criteria.

The dam safety service was initiated in the CWC in around 1980. This service brought to the notice of all state governments the need to review design floods to ensure the safety of the dams, in general, in the light of the existing criteria for new dams. In 1985, the Indian Standards Institute finalized the draft of a standard entitled 'Guidelines for fixing spillway capacity of dams'. This is somewhat more comprehensive than the existing CWC criteria in regard to storage dams. The ISI guidelines state that the standard is for constructing new dams and is not for deciding the adequacy of old structures. Thus, there was a need for separate guidelines for deciding adequacy of old structures.

Hence, the CWC came up with the suggestion that there shall be two separate guidelines, one for existing dams and another for new dams. However, the two guidelines would not differ in the choice of inflow design floods. They may differ marginally with regard to the freeboard, clearances and safety factors. Thus, although two guidelines would exist, these would not differ radically as far has hydrological safety guidelines were concerned.

According to the report on dam safety procedures prepared by the CWC, taking a decision on safety review would call for an interdisciplinary effort. 'It would be better if the existing dam can conform to the present ISI code with modifications thereof, if any. However, more often than not, it is found that they do not conform strictly to the codal provisions. The structural safety of the dam shall be assessed taking into account the latest state-of-the-art. Each case has, therefore, to be decided on its own merit and precedents shall have no place in such a review. In a situation where safety review indicates that the existing dams do not conform to the present standards, it would be obligatory on the part of the owner to plan disaster preparedness, including dam break model studies to meet the unlikely event of a failure indicating therein the possible inundation downstream of the reservoir consequent to dam break. Such dam break model studies shall take into account the deficiencies noticed in the review of the structure,' the report said.

*

As the Mullaperiyar dam was designed and built close to a hundred years before the Dam Safety Organization came out with their guidelines on dams and the Bureau of Indian Standards fixed the standard for dams in the country, it is obvious that it does not meet many of the guidelines prescribed by the DSO or even the Bureau of Indian Standards and ICOLD.

At the time of its construction, the length of the dam was 1,200 feet with a full reservoir level, that is, the maximum level up to which the water can be stored in the reservoir of the Periyar dam, fixed at 152 feet. The maximum water level (MWL), that is, the highest point up to which the dam can safely take inflow of water without the danger of toppling or crashing down, was fixed at 155 feet. Beyond the MWL, the parapet of the dam stood at 158 feet.

As per the original design, the two smaller dams on either side of the main dam were constructed at a height of 144 feet as a flood escape mechanism. The length of the baby dam and the earthen dam on the left side of the main dam was approximately 420 feet (210 + 210), while a similar spillway was present in the right side of the dam measuring around 480-odd feet, thus ensuring that a total spillway area of 900 feet was available on either side at a height of 144 feet. The flood escape saddles were designed in such a way that approximately 1.2 lakh cusecs of water could flow through these 900-odd feet of saddle, thus retaining the water level to not exceed beyond 155 feet in the dam, which is the MWL, thereby ensuring that the flood waters would not topple the dam at any rate.

The PMF for any dam is the maximum inflow that could occur in the reservoir of the dam within a stipulated period of time (usually over a period of two days by the hour), and the spillway capacity is determined after studying the PMF for the dam to ensure that the water flowing into the reservoir does not topple the construction by flowing over it.

In the case of the Mullaperiyar dam, the PMF was designed at 2.12 lakh cusecs using various calculations acceptable at the time. Hence, the spillway capacity of 1.2 lakh cusecs at FRL of 152 feet was designed to ensure that water inflow even during the maximum flood seasons did not threaten the structure of the dam. The maximum rainfall for the region which decides the PMF was decided using the historical data for the last several years of rainfall statistics collected at the time.

However, after certain defects were detected in the saddle spillway design several years after the dam was constructed, the engineers decided to put up ten floodgates on the left bank at a level of 136 feet. Hence, the saddle spillway was broken down from 144 feet to 136 feet and gates with regulators were erected that could be operated on a need basis. According to TN PWD engineers, the spillway capacity of these gates was fixed at 86,000 cusecs.

*

The dam as it is today is not what Pennycuick had constructed during the late nineteenth century. When the Kerala government took up the safety issue of Mullaperiyar with the CWC in 1979, dam safety experts from the commission checked the safety standards of the dam and suggested strengthening techniques that, according to them, would make the dam as good as a new one.

As a result, the Mullaperiyar went through a major strengthening exercise since 1979 for a period of around twenty years. A high level meeting of officials from Kerala and TN was held with the chairman of the CWC when several measures were proposed for strengthening the dam to meet modern safety standards.

The meeting took place on 25 November 1979 and was presided over by Dr K. C. Thomas, then chairman of the CWC. The strengthening measures suggested during the meeting were split into three phases: (i) emergency measures to be completed as soon as possible; (ii) medium-term measures; and (iii) long-term measures to be done over a period of time to ensure that the dam safety standards would be raised to modern levels.

One of the first measures that were implemented was the creation of three additional vents to increase the spillway capacity. Three additional vents with 40 feet width at a height of 136 feet along with the existing ten vents for flood discharge were constructed, and with the additional vents, the combined spillway capacity of the Mullaperiyar dam once again rose to 1.22 lakh cusecs.

*

The water level in the dam was immediately reduced to 136 feet from 152 feet, as per the CWC directive, and the shutters for the spillway were expected to remain open until all the strengthening measures were completed and the water level in Mullaperiyar could be increased back to 152 feet. Reinforced concrete capping was also provided from the top of the dam after knocking off one metre from the top of the dam to ensure that the construction was strengthened. The designs and drawings were finalized by the TN government after obtaining the counter signature of the chief engineer of the Kerala government. The work was sanctioned for ₹131 lakh and was completed by March 1981.

After the short-term measures were successfully completed, the TN PWD engineers took up the task of providing cable anchoring to the dam to increase its resistance to earthquakes as part of the medium-term measures. Cable anchoring was also successfully completed across the 1,200 feet of the dam using cables that were dug deep into the rocks to provide a kind of support for the wall of the dam. The cable anchoring took several years to be completed during which the TN PWD engineers claimed that they encountered stiff resistance from the Kerala state government employees who hampered the progress of their work. The structural design and drawings were completed, and the project was sanctioned for ₹304 lakh and completed by the end of the late 1990s.

As part of the long-term measures suggested by CWC, a joint team of engineers from TN and Kerala would explore the possibility of locating a new dam within a month's time as an alternative to long-term measures for strengthening of the existing dam. While the TN PWD officials claim that a joint exploration did take place and the dam was found unsuitable, the Kerala Irrigation department engineers point out that the initiative was sidelined by TN and never taken up seriously.

As far as strengthening of the dam was concerned, the commission ordered that the existing dam be backed with reinforced concrete on the rear face. 'Kerala State Electricity Board was to take immediate steps to provide power supply to the dam and investigate an approach road from Vandiperiyar–Pamba Road in the downstream of the Periyar Dam along the shortest route. For the implementation of the emergency measures, neither power supply nor road should be a precondition. Provision of both were preconditions for long-term measures,' the CWC recommendation report said.

Besides, wireless facility was to be provided by the TN government as a permanent arrangement. The Kerala government was asked to hook up their facility to this facility or, if that was not possible, a separate wireless set was to be provided at the dam site for communicating information regularly. For monitoring the reservoir level, flood discharge, etc., an assistant engineer was to be posted from the Kerala side at Thekkady, and monitoring of work of emergency measures was to be done at regular monthly meetings between the Chief Engineer, Projects, Kerala and the Chief Engineer, Irrigation, TN.

After almost all the strengthening measures were completed barring a few minor ones, the TN government approached the CWC for raising the water level to 142 feet, which is when the row began.

*

The most recent investigation that was conducted on the safety of the Mullaperiyar dam was by the Empowered Committee (EC) appointed by the Supreme Court of India after the case went back to the courts in 2008–09. The EC, headed by Justice A.S. Anand, comprising legal experts Justice K.T. Thomas and Justice Dr A.R. Lakshmanan and a panel of technical experts Dr C.D. Thatte and Dr D.K. Mehta, completed its investigations and delivered a report on the stability of the dam in May 2012.

The EC conducted its investigations from both a legal and technical point of view on the questions of stability and alleged breach of contract between TN and Kerala. It also studied the proposal for a new dam and came out with its conclusions.

The focus of the detailed investigation was to address: (i) the grievance of TN of not being permitted to raise the FRL from 136 feet to 142 feet and beyond, as per the SC verdict of 2006; and (ii) the concerns of Kerala about the safety of the dam and a fear that the dam had outlived its utility and is no longer safe despite the strengthening of the construction over the last few decades. The EC got its tests and investigations conducted through the three apex organizations of the Ministry of Water Resources, the Indian Meteorological Department and the Geological Survey of India and a few other specialized agencies whenever needed.

The inquiry was conducted basically on four main factors based on the submissions made by both the states. These factors are: (i) hydrological safety; (ii) structural safety; (iii) seismic safety; and (iv) the need for a new dam.

As far as the hydrological safety of the dam was concerned, the EC certified the Mullaperiyar dam was found to be safe for the water level to be raised to 142 feet from 136 feet. One of the most fiercely debated points by the Kerala government in terms of hydrology was that of the calculation of the PMF. The EC concluded that the PMF with a peak flow of 2.12 lakh cusecs was acceptable, and also concurred that the PMF could be routed over the reservoir FRL of 142 feet to safely pass over the dam spillway with thirteen operative gates, resulting in a peak outflow of 1.43 lakh cusecs raising the MWL to an elevation of 153.47 feet transiently. 'Even for the test case of one gate remaining inoperative, the MWL raises to an elevation of 154.10 feet when PMF impinges the reservoir at FRL 142 feet,' the committee members said.

EC also concluded that both the main and baby dams were structurally safe even if the FRL were to be restored at 152 feet. It suggested several maintenance measures to be carried out in a timebound manner including treatment of upstream surface, reaming of drainage holes, instrumentation, periodic monitoring, analysis and leading away the seepage from the toe of the dam towards downstream, geodetic reaffirmation, etc. The EC also said that the dam body should be grouted with a properly designed grout mix of fine cement/suitable chemical/epoxy/polymer, according to expert advice, so that its safety was ensured.

In terms of seismic safety, the EC concluded that the Mullaperiyar dam was seismically safe for an FRL of 152 feet and MWL of 155 feet for the identified seismic design parameters with acceleration time histories under 2D FEM analysis. 'The strength and other properties of dam material presently available indicate ample reserve against the likely stresses/impacts assessed under this analysis. In addition, reserve strength of cable anchors makes the dam further safe,' the committee said in its report.

The suspicion about existence of a geological fault in the baby dam foundation was ruled out. The recent earthquake in the dam area was considered of no consequence to the seismic safety. The EC also made several observations regarding Kerala's proposal for a new dam, which will be discussed in a later chapter on the possible solution to this controversy.

*

CWC Agrees, IITs Disagree

Although there has been a general perception among the people, technocrats and politicians in Kerala that the Mullaperiyar dam is unsafe for several decades, the state machinery did not have any concrete measure of how safe or unsafe is the dam until 2006, when the SC verdict against the state's stand in the issue woke up the people.

'It was when we actually realized the fact that we did not have any material to back our claim,' said a member of the Mullaperiyar Dam Cell formed by the Kerala government to provide technical background to the lawyers fighting the case. I refrain from divulging the engineer's name in this discussion considering the sensitivity of the issue and to ensure that the information that he shared reaches the reader as it is. Hence, let's call him J.

I met this senior engineer J during late May 2012 at a time when the EC had already delivered its report, and the Mullaperiyar cell members were busy preparing enough ammunition to fight tooth and nail when the case reaches trial stage.

'It is going to be a really difficult task for us,' J said. 'To counter a report filed by some of the top engineers of the country appointed by the Supreme Court and to prove that the dam is unsafe is an insurmountable task. But we have enough proof to show that the entire process of investigation had several flaws,' he added.

Beyond the banal and most common claims made in every public forum that the Periyar dam is unsafe as it is constructed using lime surkhi and that the dam is over a hundred years old, complex technical data has been studied and collected by both states' engineers to prove their point. A perusal of both leads one to question the veracity of some of the top institutions in the country, such as the IITs and Anna University, who have produced differing results after conducting similar studies.

'One of our strongest claims that the dam is unsafe concerns the hydrological safety of Mullaperiyar dam,' said J. 'Most dam disasters across the world have happened due to lack of adequate hydrological safety measures present in the dam and Mullaperiyar is a classic example of how hydrological safety is ignored.'

Before getting into the details, or as minimal details as possible to refrain from boring the reader, yet keeping him or her informed on the safety features, a little recap on the hydrological safety aspects of Mullaperiyar dam.

As it stands today, the dam has a FRL of 136 feet and a MWL of 155 feet, as mentioned in the lease deed, with a spillway capacity of 1.22 lakh cusecs of water. Now, what this means is that while the full reservoir is only 136 feet, the dam can withhold water upto 155 feet beyond which it will topple or break. The spillway, that is, the flood discharge outlet from the dam, is designed to flush out 1.22 lakh cuses of water only.

As mentioned earlier, spillways are traditionally designed keeping in mind the maximum possible flood in and around the dam catchment as calculated by the meteorological department and other agencies. Hence, in the event of a heavy flood that results in huge quantity of inflow in the dam, the spillway should have adequate facility to flush out that flood and prevent water from overflowing the dam.

The most important factor in deciding the spillway capacity of any dam is, thus, the PMF that could be predicted in the region. By predicting the maximum possible flood, engineers arrive at the inflow quantity of water that is possible to flow in the case of the worst flood and accordingly design the spillway.

'Thus, calculation of the PMF is one of the most fiercely contested issues in designing the safety of the Mullaperiyar dam. While the Central Water Commission has arrived at a PMF of 2.12 lakh cusecs, independent studies conducted by the civil engineering department of IIT Delhi concluded that the PMF is much higher and the present spillway is inadequate to flush out the water resulting in a possible toppling of the dam,' J told me.

*

Between 2006 and 2008, a team of engineers from the Civil Engineering Department of IIT Delhi, headed by Professor A.K. Gosain, conducted indepth studies based on latest available technologies and data and came up with results that were startlingly different from those arrived by the CWC.

Mullaperiyar cell members in Kerala point out that while the CWC studies were dated and used rudimentary techniques, the study conducted by IIT was state-of-the-art and most reliable. 'The IIT engineers have concluded that a flood of a much higher potential is possible in the Mullaperiyar catchment area based on the meteorological data they received from the Indian Institute on Tropical Meteorology, and have concluded that a PMF of 2.91 lakh cusecs is possible in the surroundings of the dam,' J said. 'For an inflow of 2.91 lakh cusec (cubic feet per second), the dam would definitely overflow even if the maximum water stored in Mullaperiyar is just 136 feet,' he added.

J pointed out that the report also concluded that as per the Bureau of Indian Standards for Dams, the spillway capacity had to be designed in such a way that the dam was safe even if 10 per cent of the spillway gates failed to operate, as is quite common during heavy flood situations. 'Mullaperiyar spillway has thirteen gates and even if all thirteen were open, even then a flood with a PMF of 2.91 lakh cusecs would possibly topple the dam. If we were to go by Bureau of Indian Standards, then we should consider only twelve flood gates operative, and the spillway capacity would be sufficiently reduced which could increase the risk further,' he said.

According to the Kerala-based engineers, the CWC has conveniently ignored this study and has gone ahead with their calculation that a flood with a discharge capacity of only 2.12 lakh cusecs was possible in the region.

'It is one of the biggest errors conducted and we will fight it in court,' J said. 'Even the CWC has proof with it that shows that during the year 1945, there were heavy floods in Idukki district causing an inflow of 2.56 lakh cusecs of water. Fortunately, nothing happened during that period. When designing such critical installations as dams or nuclear power plants, one has to consider the worst possible scenario. However, our experts in the central government have somehow ignored this data only to prove that their earlier claim (in 2006) that the dam is safe remains true even this time.'

*

Between 2007 and 2008, the Kerala government had assigned the Department of Earthquake Engineering in IIT Roorkee to conduct a study on the seismic stability of the dam and the seismic threats faced by the region in detail.

The team from IIT Roorkee, led by Professor D.K. Paul and his team, conducted detailed studies of the region and concluded that the dam would not withstand an earthquake of a reasonable magnitude that was probable in the Idukki region.

In his report, Professor D.K. Paul said that the Mullaperiyar site was located in Seismic Zone III as per the seismic zoning map of India incorporated in the Indian Standard Criteria for Earthquake Resistant Design of Structures. 'The tectonic features near to the site are the Periyar fault, Ottapalam Kuttampzhua Fault and the Kattagudi Kokkal Palani fault besides several other minor faults and lineaments,' the report said.

According to Prof. D.K. Paul and his team, historically and instrumentally recorded data on earthquakes showed that the area of Mullaperiyar and its neighbourhood was prone to earthquakes of slight to moderate intensity.

'The prominent among them are the earthquake of 1 April 1843 with a magnitude of 6.0 on Richter Scale that took place at Bellary in Karnatakka and the Coimbatore earthquake of 8 February 1900 also of approximately the same intensity that was also felt along the region. Less intense earthquakes were reported during 7–8

June 1988 in Kottayam, Idukki and Ernakulam districts of Kerala with a magnitude ranging from 3.5 to 4.5 on Richter Scale and again on 2 December 1994 with an intensity of 3.8. Some of the most recent earthquakes include the one that took place on 12 December 2000 with a magnitude of 5.0 on Richter Scale which was widely felt across Kerala,' the report said, indicating that the region was indeed prone to quakes.

Before getting into the details of the study and its results, let us go over some basics of measuring earthquake intensities. While Richter and moment magnitude scales are commonly used to measure the intensity of earthquakes, the intensity of a quake when concerned with its impact on buildings and other structures is measured using the peak ground accelerator (PGA).

PGA is basically a measure of the earthquake-related acceleration on the ground and is an important input in earthquake engineering. It gives a measure of how hard the earth shakes in a particular area during the quake. PGAs could either be arrived at mathematically in hypothetical situations or can actually be measured using the instrument accelerograph.

The seismic hazard assessment at the dam site was arrived at based on the estimation of maximum considered earthquake (MCE). MCE is defined as the earthquake that can cause the most severe ground motion capable of being produced at the site under the currently known seismotectonic framework. It is a rational and believable event that can be supported by all known geological and seismological data. It is determined by judgement based on maximum earthquake that a tectonic region can produce considering the geological evidence on past movement and the recorded seismic history of the area.

During the study, the seismic hazard assessment had been carried out by the deterministic as well as the probabilistic approach. The report said that the safety of the dam had to be checked for MCE conditions for the maximum PGA value arrived at the site. Based on the deterministic approach, the MCE conditions were estimated at 0.16g at, the site while using the probabilistic approach led to a PGA value of 0.21g under MCA conditions with

2 per cent exceedance in fifty years. The report conservatively used the probabilistic approach and fixed the PGA at 0.21g for checking the safety of the dam.

The second phase of the study was to do 2D plain stress linear dynamic finite element model (FEM) analyses of the main dam and the baby dam under MCE condition. The objective of this test was to assess the safety of the dam by carrying out 2D plain stress static and dynamic analysis of the Mullaperiyar and baby dam sections, considering dam-foundation interaction using FEM in an integrated manner under MCE condition. A 3D block analysis was also carried out.

Based on the series of studies, Prof. D.K. Paul and his team concluded that the finite element analysis of Mullaperiyar dam subjected to static and seismic loading showed tensile stresses at the heel of dam-foundation interface.

'Based on the analysis, both the main Mullaperiyar dam and baby dam are likely to undergo damage which may lead to failure under static plus earthquake condition and therefore needs serious attention,' the report stated.

It also concluded that most of the values adopted for material properties were based on the tests conducted some twenty–twenty five years back. 'During this period this dam has definitely gone through considerable deterioration due to ageing and weathering. As such the assumed parameters may be naturally higher than the in situ condition. Proper assessment of existing material properties is very important for the safety assessment. It is therefore recommended to carry out further testing on the dam and foundation materials,' the report said.

The findings of these independent organizations gave more teeth to Kerala's claim that the dam was unsafe and a new dam was needed. All these documents and reports were presented for the most recent study done by the EC constituted by the Supreme Court of India to find out if the Mullaperiyar dam was safe and will be fiercely debated as the case goes for trial.

*

Unlike hydrological safety and seismic safety concerns that could be determined without having access to the maintenance records of the dam by conducting separate studies, the structural safety of the Mullaperiyar dam could not be independently ascertained by the Kerala engineers as all the records and maintenance documents were available with the TN PWD engineers. 'We did not have any access to them despite requesting for the same. One of the biggest issues in this debate is the lack of transparency by Tamil Nadu PWD,' J told me.

Some of the main concerns regarding the structure of the dam that have been raised before the SC by Kerala include the argument that since Mullaperiyar dam is a composite gravity dam constructed of lime surkhi mortar and lime surkhi concrete, the inner core of the dam which constitutes 62 per cent of the total volume admittedly consists of lime surkhi concrete.

'Hence, the Mullaperiyar dam is presumed to be unfit for storage of water after the lapse of more than a century. Such a conclusion is evident from the heavy leaching of lime. The Mullaperiyar dam has lost as much as 30.48 tonnes per annum, totalling to about 3,412 tonnes,' he said.

In their response to the SC, members of the Mullaperiyar Cell in Kerala had pointed out that the density of the materials used in the dam had gradually gone down from 150 lbs/cft considered in 1895 to 135 lbs/cft in 1986. Compressive strength of the material used in the Mullaperiyar dam was 43 kg/cm^2 to 53.8 kg/cm^2 at the time of construction of the dam. 'However, the said compressive strength came down to 35 kg/cm^2 in 1986 and further to 28.55 kg/cm^2 in 2000. This gradual reduction testifies structural degradation of the Mullaperiyar Dam,' the members of the cell said in response to the safety report. 'State of Tamil Nadu has neither conducted any investigation nor made available any core samples in the recent past from the Main Mullaperiyar Dam to establish the present quality of the material,' the report added.

The Mullaperiyar dam stands on rock which has not been

investigated and treated at all. Also, the foundation is not watertight because two of the boreholes numbered 35 and 39 absorbed 59 to 247 bags of cement during grouting, and boreholes numbered 21 and 23 absorbed 247 and 202 bags of cement, respectively, during grouting.

They also pointed out that the Geological Survey of India (GSI) conducted a geological mapping of the Mullaperiyar dam area as part of the Dam Safety Assurance and Rehabilitation 21 Project. In its report, GSI advocated the study of the seismogenic capabilities of lineaments by instrumentation and monitoring of the Mullaperiyar region.

'But this has not been complied with by Tamil Nadu. Regarding the baby dam, it was not considered for adopting as a spillway by modification in 1980 because Tamil Nadu itself thought that the foundation is weak,' the members said in the report.

*

Some of the questions that the vastly differing technical studies conducted by people considered as experts in their chosen field raised are regarding the credibility of several institutions such as the CWC, the IITs and the reputed Anna University in TN.

J points out that there have been serious differences in the results produced by the CWC and the IIT in terms of both hydrological safety and seismic safety. 'While IIT says that the Probable Maximum Flood is 2.91 lakh cusecs, CWC says it is 2.12 lakh. Even in terms of Peak Ground Accelerator (PGA) values, IIT Roorkee's Earthquake Engineering department, one of the most prominent of its kind in the world, comes out with a finding that the PGA is 0.21g, but the CWC engineers finds the testing and the results inadequate and instead relies on other data. So who are we to believe then?' he asks.

He points out that both states should jointly appoint a third party agency, preferably international, to conduct independent tests in the most transparent manner. 'One of the biggest issues in this entire controversy is the lack of transparency,' J says.

*

While Kerala's prime contention has been to prove that the dam is unsafe based on independent tests and studies conducted by IIT Roorkee, IIT Delhi and other organizations, TN's contention has been to prove that the dam is as good as new, and that it can hold up to 152 feet of water as the strengthening measures suggested by the CWC in 1979 have all been, by and large, completed.

Members of the TN PWD Senior Engineers' Association who have been taking up the cause of the farmers and have been providing the people living in the regions benefitted by the dam with technical information, claim that now that the dam is safe, there is no reason to continue limiting the FRL to 136 feet.

The association claims that the only real way to test the dam for safety and decide on an accepted FRL would be would be to conduct a real-time stress test. The CWC and the EC have already done stress tests on the dam using the 2D finite element method (FEM) and the 3D FEM tests, and have theoretical data on the amount of stress that vulnerable points of the dam could hold.

A retired chief engineer with the PWD, who had headed the Madurai region which includes the Mullaperiyar dam during the early 2000s, states that neither the CWC nor the Kerala government had requested for a real-time test as it would prove that their theories on the dam's safety were wrong.

According to him, the present FRL could be gradually raised by about a foot every day by blocking the tunnel through which water flowed into TN for a specific period of time. 'Just about ten days of blocking the tunnel would be sufficient to perform the actual stress tests,' he said.

By stopping the outflow of water, the dam's FRL would gradually increase beyond 136 feet, and at every foot of rise in water level; the stress on the dam construction could be measured. 'If the dam is able to safely take a stress of 142 feet without showing distress, then we can fix that as the present FRL. If the dam can take an FRL of 152 feet, as originally designed by Pennycuick, then why should we reduce the FRL without any scientific basis?' asks Er. Vijayakumar.

He pointed out that strain gauges were available to test the strain endured by various parts of the dam and they could be burrowed to find out the actual safe FRL levels. 'However, despite the availability of such techniques, neither the CWC nor the TN government has put forward this demand. The best way to test the Mullaperiyar is to burrow strain gauges and increase the water level,' he said.

It has to be noted that the water level at the Mullaperiyar dam has never been increased beyond 136 feet since 1970. The Government of TN has spent several crores of rupees to strengthen the dam, and the state engineers feel that it is a shame to not have tested its strength by increasing the water level after taking the necessary precautions.

A real-time stress test conducted by an independent international agency, using strain gauges and by gradually raising the water level, could go a long way in dispelling several myths about the dam safety under present conditions. Yet, it has never even been mooted so far.

*

5
The Background

FOR THE CASUAL visitor, Palarpatti is like any of the hundreds of tiny hamlets that dot the lush green landscape between Cumbum and Theni in southern TN. Anchored by a meandering river with fertile farmlands on both sides and a stretch of modest homes cuddled together in the centre of the settlement, the locals claim that around 650 families, most of whom are farmers, live here.

But Palarpatti is unlike any other village in TN or anywhere else in the country, for it is probably the only Indian village where a British engineer is revered as God and is worshipped even so many decades after his death.

Colonel Pennycuick, the British engineer who built the Mullaperiyar dam between 1887 and 1895, is that man. Huge potraits of Pennycuick have been framed and hung in every home in the village, and the residents of Palarpatti take pride in celebrating the legacy of this foreigner. Even the community hall in the village has been named Pennycuick Mandapam after him.

Every year on 14 January, even as everyone in TN celebrates Pongal, the harvest festival also known as Tamizhar Thirunal, residents of Palarpatti celebrate Pennycuick's birthday which also happens to be on the same day.

All the villagers and their family members assemble at the community hall in Palarpatti and cook pongal, a dish made of rice and daal, in the village centre with a huge portrait of the Britisher whose thick brooding moustache and probing eyes oversee the festivities. Later, the villagers conduct games such as kabbadi, kolam (rangoli) drawing, cockfights, etc., to commemorate the occasion, and distribute trophies and prizes in the name of Pennycuick.

Even children in Pallarpatti and surrounding areas in Theni district know about the British Engineer J. Pennycuick (1841–1911) who built the Mullaperiyar dam, and can narrate interesting anecdotes of his biography that have been embellished over the years with several interesting fictional elements adding spice to the story.

In a country where successful movie stars, politicians, sportspersons and, in rare cases, physicians are seen as demigods, British engineer J. Pennycuick is probably the only engineer who is considered God and a hero of the people.

Residents of Palarpatti have enough reasons to revere Col Pennycuick as a god. They believe he transformed their lives forever by undertaking the unfathomable feat of constructing the Mullaperiyar dam.

'This region is dry due to scant rainfall and the absence of any perennial source of water. Although the Surliyar has been flowing here for centuries, it is highly seasonal and does not provide enough water for farming activities. At best, the Surliyar could be used to irrigate one crop,' says Deivam, an elderly farmer from Palarpatti.

'In the absence of any steady farming activity, our ancestors had no choice but to resort to other activities to make a living which are now considered illegal. Brewing arrack and waylaying travellers was the primary occupation of this region during periods of drought. And droughts were pretty regular here,' he says.

Today, even the oldest men and women in Palarpatti do not recall ever having to struggle for a meal. Ever since they have been born, the farmlands around their village have been producing two yields in a year, providing them sufficient foodgrains to fulfil their needs as well as to sell outside. 'We have been told by our parents and grandparents of how this village and other areas nearby have been transformed after the dam was built. Hence, we consider this man Pennycuick as our hero and saviour,' said Andi, an elderly person from Palarpatti.

Not just in Palarpatti but in several other villages surrounding Bodi such as Surulipatti, Narayanathevanpatti, Kullagoundanpatti, etc., Pennycuick is a well-known figure and the public here pay homage to the man year after year.

John Pennycuick was born in Pune in 1841 to Brigadier General J. Pennycuick and Sarah. After completing his education in England, Pennycuick returned to India in 1860 and worked in the Madras Public Works Department for several years before retiring from service as the chief engineer in 1896, a year after the Mullaperiyar dam was completed.

Incidentally, Pennycuick had not always been a legendary hero in the villages of Theni district. The controversy that has been brewing over the past thirty years has probably resurrected the dead Colonel from his grave and has given him a place in the hearts of the public here.

'We have started celebrating his birthday and legacy for a little more than a decade now. Until then, most people were not aware of this man or his contribution to our state,' says Mani, a 32-year-old man from Palarpatti. 'Some years ago, a man from this area went to a college in the city and learnt about Pennycuick. Later, he wrote a booklet about Pennycuick and distributed it among us to educate us on the Mullaiperiyar dam and the man who had constructed the dam,' he said.

*

Irrespective of whether this adoration in Palarpatti and surrounding areas for Pennycuick, whose accomplishments are much lesser when compared to legendary engineers like Sir Arthur Cotton, is motivated or not, there is no denial of the fact that his contribution to the village that has no perennial river and harsh climatic conditions is immense, as is obvious today.

Scant rainfall, undependable water sources and recurring famines and periods of drought are not just part of the history and folklore of Palarpatti but of the entire region extending across

Madurai, Theni, Dindigul, Sivaganga and even some areas in Ramnad district. The significance of the Mullaperiyar dam can never be understood unless one understands life in this region before there was a dam.

The only major river that flows through this part of TN is the Vaigai. This river is not perennial and is considered one of the medium-sized rivers in the country. The Vaigai basin is one of the thirty-four river basins in TN and ranks fourth in its water potential.

The Vaigai basin extends up to 7,030 square kilometres and is spread over the districts of Dindigul, Madurai, Sivagangai and Ramnad. The leaf-shaped basin is bounded by the Western Ghats on the west, the Kottaikara Aru basin on the north and Uttarakosai Mangai Aru basin in the south. It narrows down towards the Bay of Bengal in the east where it disappears in the vast coastal sands. Madurai is the only major city that lies in the Vaigai basin. The other towns of importance that lie along the basin are Ramanthapuram, Periyakulam, Usilampatti, Thirumangalam, Manamadurai and Paramakudi.

The Vaigai originates in the Varshanad ranges of the Western Ghats, draining the eastern slopes, and runs north for around 65 kilometres to be joined by its tributary Surliyar on its left, and thereafter turns east and runs east, southeast past Madurai city for another 80–85 kilometres before it shrinks in size having been abstracted to fill in a large number of tanks along its way. Towards the end, after it empties most of its water in the Ramnad big tank, the river gets lost in the east coast. According to TN PWD engineers, the river has not created either a delta or even a piercing defined course to fall into the sea. The absence of a delta or a merging with the sea clearly explains its low water potential and the maximum utilization of its water. Some of the other tributaries that join the Vaigai along its way are Theniyar, Varahanadhi, Manjalar, Uppar, etc., besides a few more that join the river along its course.

As the Vaigai basin lies on the leeward side of the Western Ghats, it misses the benefits of the dependable southwest monsoon and receives its rainfall from the northeast monsoon which is less dependable and highly unpredictable. Rain during the northeast monsoon is usually accompanied by depressions and storms generating heavy floods in the river, and during a bad monsoon season causes severe drought in the region.

Considering the unpredictable nature of the Vaigai, even centuries ago the administrators of the region had already developed a solid network of tanks and channels that drew water from the Vaigai and diverted it for irrigation. However, many of these tanks were so silted that their potential was not fully utilized and could hold only half the quantity of water that they were designed to hold. The scarcity of water also led to several disputes among peasants as they diverted water from these tanks to irrigate their fields.

In a report compiled by the then district engineer Major Ryves dated 7 August 1867 regarding the investigation into the failure of the Vaigai to irrigate more than 22,000 acres of land that was then being irrigated by the river, he states that while the tanks were extremely essential considering the nature of the Vaigai known for its frequent floods, the water received into the channels only bore a small proportion to the rainfall in the region, and that much of the water evaporated due to the shallow nature of the tanks.

In his report, Major Ryves says: 'The supply of water obtained from the Vaigai is so precarious and scanty that even in good years the paddy crop barely covers 22,000 acres annually, although the existing tanks and channels command land enough and have sufficient hydraulic capacity for the irrigation of double that extent of crop, if only a sufficient supply of water, delivered at a regular moderate rate, be ensured.

'Under these circumstances it is not surprising that agricultural operations are seldom rewarded by a good crop. If the ryots (farmers) are so fortunate as to not find the ground as hard as brick at the ploughing season, the chances are that rain necessary to bring the crop to maturity will not fall at the expected time.

'And the cultivation of wet crops is hardly a less precarious business, failure being attended with greater loss, and success gained at much expense of labour and money on raising water from wells and pools.

'Almost every alternate season is one of scarcity in this taluk, and when an exceptionally dry year occurs, there is severe distress, and the population is thinned by death and emigration. In 1861–82 and again last year, it suffered severely in this way.'

*

While famines occurred on a regular basis across the country, the Madurai region was especially witnessing frequent periods of drought due to the nature of the Vaigai. Since the river received most of its rainfall from the northeast monsoon in the form of thunderstorms and cyclones, the flow was so heavy during the rainy season that most of the water could not be utilized for agriculture, as Major Ryves had mentioned in his report.

Although one could not directly associate a famine with a drought period, as various studies have now indicated, it is safe to assume that a recurring or a prolonged drought period could be one of the main reasons to trigger a famine. The Great Famine of 1876–78 that affected south and southwestern India and wiped out anywhere between five million to nine million persons from the south alone, according to historians, had a great impact on Madurai too.

In the book *History of the Periyar Project* by A.T. Mackenzie, written in 1897, the author, who served as chief engineer after Col Pennycuick, has mentioned that the British government had spent ₹432,170 on famine relief during 1876–77 and another ₹792,047 on gratuitous relief in Madurai district. He also states that 'loss of revenue and life is beyond computation' during the famine days.

The British who were so keen on maintaining accounts and ensuring that every enterprise was profitable certainly did not appreciate the recurrent expenditure that they bore due to droughts and famines. At that time, the only option with the British to put an end to the famines was to ensure that the area of irrigated land was increased and that the farmers got steady supply of water to produce a good crop year after year. While the idea of diverting the water from the Periyar towards the east into Madurai had been existing for a long time, and some claim that even before the British rule, various local kings had considered diverting the waters, a serious search for an alternative water source to irrigate the farmlands in and around Madurai besides the Vaigai began only during the second half of the nineteenth century. This search led the British to the Periyar basin on the other side of the Western Ghats.

To the west of the Vaigai basin on the other side of the Western Ghats is the Periyar basin. The Periyar, as its name suggests, is one of the longest rivers in Kerala state flowing along a length of about 244 kilometres. The total catchment is 5,243 square kilometres of which about 114 square kilometres flows within the state of TN.

The Periyar rises in the Sivagiri peak of Sundaramala in the Western Ghats, around 80 kilometres south of Devikulam at an elevation of around 2,400 metres, and traverses south through steep cliffs and dense forests for about 16 kilometres where the tributary Mullayar joins it on its right at about an elevation of 850 metres. The river then turns west, cuts through the hills in a narrow deep gorge at about 11 kilometres below the Mullayar junction. The river then takes a winding course until it emerges in Vandiperiyar in Idukki district.

The river then traverses through the taluks of Peermedu and Devikulam of Kottayam districts and parts of the Ernakulam district. Besides Mullayar, several other tributaries join the main river on either side, including Kattapanayar on the right, Cheruthoni on the left and the Edamalayar. As the river reaches Alwaye, it divides itself into two branches. The principal branch flowing north west joins the Chalakudi River and then expands itself as a broad sheet of water at Munambham. The other branch taking a southern course breaks into a number of distributaries and falls into the Vembanad Lake at Varapuzha before merging with the Arabian Sea.

Although the main catchment of the river lies inside dense, impenetrable and uninhabited forests, as a result of which there was no exact measure of the actual rainfall in the catchment areas of the river, the early investigators were confident that the southwest monsoon brought intense rainfall in the region and the river would provide a reasonable amount of water to be able to irrigate large extents of lands in the Vaigai valley, if alone all that water could be transferred across the mountain ranges to the Vaigai basin in the east.

*

The first proposal for diverting the Periyar waters from the western slopes towards the east to Madurai was looked into during 1808, when the British investigator Sir James Caldwell visited the neighbourhood and checked out the site. However, he seems to have confined himself to a mere diversion by means of a direct cutting from the Periyar through the watershed, and finding a rise of over a hundred feet between the two basins. But he condemned the project as 'decidedly chimerical and unworthy of any further regard', to put it in his words.

But the subject was mooted again from time to time, and work even began in 1850 to build a small dam and channel for diverting a small tributary of the Periyar, the Chinna Muliyar, based on the proposal of one Captain Faber. But the work was stopped after the coolies suffered from fever and also due to the excessive wages they demanded.

After several years, the project was again revived by Major Ryves R.E., district engineer at Madurai in 1862. He spent several seasons in local investigations experiencing great difficulties from the uninhabited and inhospitable nature of the region, the incessant rain, the absence of any road network or even paths, the dense forests full of elephant grass and leeches, also from the fever that was very common during the dry months. After an investigation that lasted almost five years, Major Ryves submitted a detailed proposal for diversion of the Periyar in 1867.

He proposed the construction of an earthen dam 162 feet high across the Periyar, with an escape to meet at 142 feet above the river bed to divert the waters into the Vaigai valley by cutting through the watershed through an open channel. The channel would lead to the Surliyar River, a tributary of the Vaigai.

The main problem in this proposal that was seen to be most

difficult to solve was the control of the river flow during the construction of the dam. The unhealthiness of the country limited the working season to the period between June and February. But the high discharge that the river would carry during the southwest and the northeast monsoons still limited the time available to work to a possible thirty days between August and September and another three and a half months from December.

Major Ryves estimated the cost of the project to be ₹17,49,000 and he also recognized that it would be a gigantic task bristling with several uncertainties and problems of the like not so far tackled anywhere.

'But those were the days when the name of Sir Arthur Cotton was fresh in the land, and the Upper Anicut and Lower Anicut across Cauvery were built and the Godavari and Krishna Irrigation projects were successfully completed. This brought into play a tradition of the Madras Engineers to shrink away from no task, however gigantic, and to make light of all obstacles that could come in their way,' says A. Mohanakrishnan, one of the most senior hydrologists in the country and an expert on Mullaperiyar.

However, the details of the scheme as contemplated by Major Ryves came in for considerable criticism and there was hesitation in accepting the proposal.

The next round of investigations commenced in 1870 and was taken up by R. Smith. He felt that on a closer look the expense would be much greater than what had been imagined, and it would be hazardous to attempt to control the river during construction in the manner proposed by Major Ryves. He suggested a change in the site of the dam and proposed raising the earth dam for 175 feet by silting.

Although Mr Smith's proposal was generally approved, the execution of the project was opposed by the then chief engineer General Walker R.E. mainly on the ground that sufficient experience had not been gained for the silting process to justify confidence in it for a work of such magnitude. The advisability of the construction of masonry instead of silt was also mooted.

*

After the idea of constructing a masonry dam was mooted, the Government of Madras called for a report from Col Pennycuick and R. Smith on an alternative proposal for building a masonry dam. Mr Smith too agreed that a masonry structure was a better alternative which he ignored earlier only due to its high cost.

Col Pennycuick soon started working seriously on the design of a masonry dam based on Molesworth's formula, but this was also not favoured. The Government of Madras referred the whole matter to the best English opinion and forwarded the estimates to the Government of India with their recommendation. The British government, however, considered that the experience of engineers in India in the construction of irrigation works must far exceed that of engineers of any other country in the world, and they offered to appoint a committee of high standing, selected from Bengal, to which an officer of the Madras PWD having complete knowledge of the locality and of the details of the project be included.

But the project was again shelved mainly due to the severe famine of 1876–78 and also because the investigations did not look into the probable returns of revenue. No further action of practical nature was taken during the next six years, but there was a great deal of discussion during the course of which the project attained a definite and less debatable shape. Finally, the whole of the papers were handed over to Col Pennycuick, who was directed by an order dated 8 May 1882 to be relieved of other duties with a view to his undertaking the revision of plans and estimates for the entire project.

Col Pennycuick submitted the entire project report with detailed estimates within a year and they were eventually sanctioned. Even at the time of framing the proposal documents, Pennycuick put in every little detail into the proposal and gave shape to the Mullaperiyar dam as it is today.

Pennycuick's proposal reduced the height of the dam to 155 feet from the bed of the river with a parapet of five feet in height and four feet in thickness. The thickness of the dam proper was to be 12 feet at the top and 115 and 3/4th feet at the bottom. He proposed it to be constructed throughout of concrete composed of 25 parts by measure of hydraulic lime (ground but not slaked), 30 of sand and 100 of broken stone. The front face is to be plastered with plaster composed of equal parts of lime and sand, the proposal document said.

The submission of the detailed plan and estimates completed the investigation of the project but there was still one obstacle to its execution. There was a disagreement between the Travancore government and the British on the terms on which the use of water and the land submerged by the reservoir should be handled.

The British government took the stand that the water was useless and likely to remain useless to Travancore for the foreseeable future, and that the land was a piece of uninhabited jungle, not of great value even for its timber and, as from the location, the woods were even impossible for the Travancore government to exploit.

The Travancore government, however, contended that the value should be appraised by its utility to the British government, which was high, since an expenditure of ₹53,00,000 was expected to bring in a return of 7 per cent per annum.

After several debates, it was agreed that the British government should pay an annual rent of ₹40,000, and that the lease should run for 999 years with the right of renewal, and that for this consideration the British government should receive a grant of the land alongside the Periyar below a contour line of 155 feet above the deepest bed of the river at the site of the dam, to the amount of 8,000 acres or thereabouts, and also an additional area not exceeding 100 acres to an unspecified level, all water flowing into the first mentioned tract, all timber growing on the said tract, and the fishing rights, with the liberty to make road through Travancore territory to the site of the works. All sovereign rights were reserved by the State of Travancore and the subsequent intricacies of civil and criminal jurisdiction, akbari rights, customs, etc., constituted a source of dissent, which lasted till the headworks were completed and until this day, more than

110 years later. If anything, the dissent has only snowballed into a huge ball of distrust for two neighbouring states.

*

The Ordeal of Constructing a Dam in the Middle of a Jungle

During the several road trips that I took to Kumily and various other parts of Idukki, Madurai and Theni districts to interview people for this book, one particular incident that took place while driving alone in Idukki district on a warm summer evening put this entire effort into perspective. Until then, I was trying to conceal my skepticism at the prospect of writing a book on an ageing dam by mindlessly running around collecting data and interviewing people.

After visiting Father Joy Nirappel, chairman of the Mullaperiyar Samara Samiti at his office in Upputhara and visiting the six-month long relay protest venue at Chappathu, I was driving back to my hotel in Kumily along a steep, winding road that runs parallel to the Periyar River Valley, with the sunset behind me. The road was almost empty except for the occasional jeep that sped past me transporting casual labourers from Theni district who worked in the cardamom estates here. On one side of the road was a deep valley at the bottom of which the Periyar, that had now been reduced to a trickle, flowed. Steep slopes covered with tall, green grass and wild plants accompanied me on the other.

I was driving past Anavilasam, one of the many tiny hamlets dotting the landscape that is identical to every other village I had passed along the road, with a teashop, a provisions shop, a barbershop and an autorickshaw stand at the centre of the village where a few men get together, smoke beedis and discuss politics after a long day's work, when a huge bug flew inside the car through the driver's side window and I lost control of the vehicle.

The sudden appearance of a giant beetle, almost the size of my palm, on the steering wheel of the car shocked me to such an extent that I almost drove into the valley. With the car perched just a few inches away from the steep slope, I applied the handbrakes and jumped out of the car. The beetle, bluish black in colour with a coarse exterior was larger than any insect I had ever seen and seemed to have sprouted out of prehistoric times. The bug that was as scared as I was kept ramming on the glass pane in the front trying to escape.

After I gained a little more confidence, I took my out my notepad and gently shoved the beetle out of the car through the open window. Later, I stepped outside, lit a cigarette and sat on a rock facing the valley to give time for my heartbeat to settle down to a steady rhythm.

As the sunlight disappeared from the valley with every passing minute, engulfing the dense jungles in darkness, I felt like I was not in Kerala but in the middle of a Jurassic Park movie somewhere deep inside an uninhabited tropical island. The deep valley surrounded by thick, impenetrable jungles; the tall trees that grew along the slopes and reached the top of the valley whose thick barks and branches seemed hundreds of years old; and the giant creepers that spread almost everywhere with leaves that were almost the size of banana leaves seemed like a perfect home for such creatures as the beetle that sneaked into my car. Within a few minutes, the noisy buzz of crickets and all other kinds of insects filled the valley, and for a minute, I pondered over how inhospitable this region must have been a hundred years ago when there was no electricity, no tar roads, no homes or any semblance of civilization of any kind anywhere in the surrounding area.

The magnitude of the effort it would have taken to construct the Mullaperiyar dam in Idukki district, Kerala would be impossible to comprehend for anyone who has not visited the thick jungles and the inhospitable terrain. A hundred and ten years ago, even bustling cities like Coimbatore were spread across just a few square kilometres with a vast stretch of wilderness all around. To take up such an endeavour and finish it successfully is nothing short of a wonder.

To imagine a few white engineers and hundreds of workers and coolies fighting their way through this impenetrable jungle to build a world-class dam cannot but be inspirational. This retelling of the construction of the dam is a blend of that overwhelming feeling with the factual data on the construction of the dam available through reports and documents.

*

The construction work began in September 1887 with the felling of a tree. Lord Connemera, then Governor of Madras, accompanied by Colonel Hasted R.E., secretary to the government, inaugurated the project by felling a tree at the site of the dam.

Clearing the dense jungles and getting the workers and staff acclimatized to the surroundings was one of the biggest challenges faced by the engineers throughout the construction period that lasted nine long years.

As with every major project taken up during the time, finding good quality workers was one of the major hurdles faced by the engineers. Thanks to the frequent droughts in Madurai and Ramanathapuram districts, labour was easily available at the time, but getting these folks used to the jungle terrain took a long time and hundreds of them died due to disease and accidents during the period of construction.

To retain the labourers and keep them from running away, only male coolies were first employed on a daily wage basis at the rate of six annas (38 paise) per diem which was a handsome pay during those days. Anyone who could read or write and had the courage to enter these deep jungles was employed as a maistry, whose main job was to organize the labour which was a difficult task at that time. The engineers even went to the extent of making advance payments and established mutually beneficial relationships to retain the good workers throughout the construction of the dam.

A vast majority of these workers stayed in coolie camps that were laid out at Thekkady along with an officers' camp on a ridge surrounded by swamps about a mile away from the Gudalur trunk road. A road was formed from the trunk road towards the dam site for part of the distance, while the remaining was covered through a footpath to the construction site. Construction of these quarters, coolie lines and other basic amenities such as first aid hospitals were some of the first constructions that came up, and all this work was completed and ready by March 1888.

According to reports filed by Col Pennycuick, a great number of coolies also came from Cumbum valley and were within reach of their homes. The high wages paid gave them the freedom to spend, which they were not accustomed to earlier, and hence, many of the coolies returned frequently to their villages and enjoyed themselves during every festival in their neighbourhood.

Most of these coolies were pledged to money lenders and farmers from whom they had received advances in cash and grain, and so, they returned to the farmlands during ploughing and harvesting times to work off their debts. The British found that their honesty was admirable and traditional and welcomed the workers back during the next season.

During the period of construction, Portuguese carpenters from Cochin and even detachments from the First and Fourth Pioneers were lent for service at the Periyar during 1889 and 1890. Whenever labour was scarce, bad and ill-organized, Pioneers were rendered into service and the officers made a welcome social addition at the site. However, the quality of their work was unequalled and they were very expensive. Occasional unpleasantness and differences occurred between the engineers and the military men which required deft handlers on both sides to prevent the skirmishes from getting serious. After 1890, the services of the Pioneers were not utilized and the labourers had by then become more regular and abundant.

Due to the harsh climatic conditions that prevailed at the site of construction, work was possible only for five months in a year for the most part, and even that was interrupted as the Periyar receives rainfall from both the monsoons.

*

One natural advantage of the site was that even though it was far and remote from human habitation, the materials of construction were available in adequate quantities and were in close proximity.

Stone, the principal constituent of the dam, found at the location, was of hard syenite variety weighing about 180 lbs per cubic foot. The syenite rock formed an excellent foundation for the dam being fairly free from cracks and fissures and remarkably homogenous. The lime used was obtained from nodular kunkur excavated from quarries near Kuruvanth at the foothills on the Madurai side. Prior to commencement of work, experiments were conducted on the strength and setting of this lime and found to be satisfactory. Although the construction works were taken up in the middle of a dense forest, the availability of timber was limited. Teak (Tectona grandis) was generally used for all purposes and black wood (Delbergia latifolia) was used when a very hard wood was required. The rest of the varieties met the fuel needs and the charcoal for lime kilns, boilers and other steam plants at the site.

The other major impediment in the smooth flow of work was the difficulty faced by the team in transporting materials and equipment to the dam site. The site is located at about 8 kilometres from the Thekkady camp on the other side of the Gudalur hills. The contours of the region were such that no mode of transport could be easily planned, designed and maintained free of breakdowns for carrying materials from the top of the hills to the main dam.

Several possible arrangements were considered and tried in parts during the construction process, and more than one mode was used in tandem to meet the needs. At least five different modes of conveyance were considered feasible during various periods of construction. The first was a metalled road linking the existing road on the watershed on which it was intended to run traction engines. The second was a similar road but shorter and with heavier gradients without traction engines. The materials could be carried by ordinary carts. The third option was a narrow gauge railway. An overhead wire ropeway was also considered, besides constructing a canal on the Mullapanjan, a small tributary of the Periyar having its head near Thekkady and running into the Periyar about a mile around the dam.

The use of a metalled road and common carts were an easy and certain option but slow and inexpensive. The advantage of this mode of transport was its simplicity and elasticity. Reports indicate that when the use of traction engines ran into difficulty, most of the material including the limestone used at the site, the grain and other supplies for workers were carried to the site on carts through the road. This mode became difficult over the years as the water level rose in the dam.

The overhead ropeway was also constructed from up the hills to Thekkady. The expense and difficulty of construction were so great that further extension was not favoured. However, heavy goods could not have been carried on it and a fairly good road as an auxiliary could not be dispensed with.

At that time, a canal appeared to be the most suitable mode of transport of all the different methods considered. A small river needing little alignment was already running in the required direction and the total length being within eight miles were the pointers that favoured this mode of conveyance. It was imagined that rock for foundation would be available within a short depth and, hence, the dams and locks could be constructed at a cheap cost. Actual transport on water surface would be really cheap and management would be within the competence of the local labour, which was always a concern for the engineers. The canal alignment was finalized as part of the proposal with a stream to be chosen.

*

The site of the dam is one of the most beautiful spots in the world. It has been and still continues to remain a haven for nature enthusiasts and wildlife lovers, particularly elephants. But deadly fevers of all kinds lurked behind the pleasant countenance of the jungle. The moderate height above sea level, vast tracts of virgin forests, the strong sun and heavy rainfall constituted favourable conditions for an active and stubborn malaria virus which was one of the greatest hindrances to the work. Annual stoppage of work was inevitable in the hot season when fever was virulent. Things did not improve even as the water rose in the lake, as the large area of vegetation that was submerged and rotting only added to the disease. The workers who were ill-fed, ill-clothed and reckless in their sanitary habits were more liable to illness due to the cold and dampness which they were unaccustomed to, and many of them whose immunity was weakened by malaria often ended up with rheumatism, dysentery and other pulmonary complaints. Hundreds of coolies died every year due to poor health and the outbreak of epidemics which were a regular occurrence.

Besides, the terror of wild animals too had to be met with. Elephants had to be constantly scared and driven off with tomtoms and firebands. Despite all the precautions taken, a man was killed by a tiger near a woodcutter's camp and another was found mauled at the doorway of his hut during this period. These incidents, too, had a telling effect on the progress of the works.

While many of these difficulties continued through the construction of the dam, some of them were eased as work progressed and the workers and engineers became familiar with the terrain and understood the conditions better.

*

The first and fundamental problem posed before any engineer who steps into a river bed to construct a dam is to find suitable means to divert the flow of the river and take up foundation excavations so that he is able to control its flow during the construction.

The problems faced with respect to the Periyar dam were unique. The dam site was so heavily overgrown with thick jungle of the most impenetrable nature that until it was totally cleared, the builders could not even get an overall view of the site, and without a clear picture of the river bed it was impossible to come up with a diversion plan. Secondly, while the depth of the lake had been taken at the general river bed level in the dam site, the masonry structure had to be raised to 144 feet above the datum where a saddle was available for the flow to be diverted, and the management of the river flow of all seasons had to be so carefully done that there was least damage to the structure that was constructed as the lake level rose.

The torrential nature of the Periyar was the next major issue, as the volume of water it carried during a great part of the year and the frequency and the strong floods from local thunderstorms even during dry months rendered a picture so varied and unpredictable that designing a satisfactory arrangement for flow diversion posed a complex problem.

But the biggest uncertainty was that the engineers had no reference to look at as no dam had been built anywhere in the world across a river so large as the Periyar at that time, combining so many unique characteristics including the compulsion of creating such a deep lake to facilitate the transbasin diversion across the ridge. This being a pioneering effort, the Periyar dam construction was later referred to by several engineers and studied deeply for use in similar such works.

The method of disposing the water of the river during construction being an important subject was carefully studied by Pennycuick (at that time) when he made his first proposals in 1882, in which he suggested provision of culverts of requisite capacity for driving tunnels at a low level, two on the left bank and one on the right. However, this proposal was rejected by the inspector general of irrigation of the Government of India who took exception to the proposal and passed severe criticism.

The strict monitoring of the progress of the work by the British government and the bureaucratic obstacles faced by Pennycuick even in taking simple decisions concerning the dam had further complicated the engineers' work, as they had to wait for approval for every decision. Although many of the decisions that were referred to the bosses in Madras and England did come through eventually, Pennycuick and his associates were forced to fight long battles to make prudent design choices as the work progressed. One such emotional letter sent by Pennycuick to his bosses for proposing the construction of a sluice at 60 feet highlights the plight.

He wrote, 'I have kept silence for upwards of ten years on the subject of the objections of the Inspector General of Irrigation, and should prefer to keep silence for ten years longer. In the present connection, it is sufficient to say that the decision of the Government of India to prohibit a low level tunnel, unfortunate though I consider it, has been accepted and loyally adhered to but that I do not consider there is any analogy between such a tunnel and the one under consideration might rightly or wrongly be urged against some of the details of the original design.' The rest of his note explained why it was necessary to change that suggestion of the Inspector General of Irrigation.

For the first two or three seasons of construction, work progressed at a snail's pace mainly due to the damage caused by the frequent floods. With the working season reduced to just five months, the construction team had to bring the work to a reasonable state before abandoning it until the next year.

By the end of 1889, work on the diversion dams for the purpose of building the canal were almost completed. Another team of workers still struggled hard to divert the river through the culverts and construct the front wall of the dam. This was a period of great difficulty and several daring and often innovative measures were taken up to overcome the hurdles posed by the river. The workers braved the harsh climatic conditions and worked day and night to ensure that all the efforts put in did not get washed away by large floods. One such brave act by the workers, aka coolies, in protecting the walls of one of the cross dams that was constructed to divert the water from getting washed away is illustrated in Pennycuick's own words.

'The cross dams even when finished had to be incessantly watched, and an emergency gang with a large quantity of earth was kept ready day and night to repair the cavities that constantly occurred. Many of the coolies were extremely good at the work, being experienced to it from childhood. A ring of them would form round a cavity pressing close leg to leg and almost excluding or at least breaking the rush of water. They were then buried up to their waists in earth by their comrades, and pulled out by the main force, when the operation was repeated until bit by bit the breach was healed. Much of this work took place at night and the exposure in water at a level of 3,000 feet above the sea in December was very trying. The coolies had, therefore, to be encouraged and assisted in every possible way and had it not been for the medicinal virtues of arrack it is difficult to see how the Periyar Dam would ever have been built. The strain on the staff was of course also very great.'

By the end of 1890, work on the main work shed got ready and the detachments from the First and Fourth Pioneers were called for their service at the Periyar. During that season, the work was again washed away on several occasions by the raging floods. During the July floods, a section of the canal locks were washed away and had to be recommenced. The river continued to attack their work through the year and a lot of damage was done during this period. However, as the main wall started rising above the bed of the river, the work became more and more steady over the years.

In 1891, the work was submerged at least five times during the southwest monsoon and 7,000 cubic feet of rubble masonry and 20,000 cubic feet of concrete were washed out, and in November, again, it was submerged four times. These floods, however, did not damage the main dam itself to any appreciable level, but often washed away piers or isolated structures which had to be replaced against the full force of the stream. Despite all these setbacks, the average monthly progress gradually increased with increased organization and experience.

By March that year, the entire work on the canal was completed and it began to be put into use for transportation only until October that year when a giant flood washed away a section of the dams. The repair and rebuilding work continued until January 1892.

One major accident that shook the engineering community that year was the sudden death of the then superintendent of works H.S. Taylor that took place on 12 October 1891. Taylor was the executive in-charge of the works from their commencement in 1887 until his death and was seen as an energetic and skilled professional whose contribution was immense.

By March 1892, that is, five years after the work commenced, the front wall of the dam was finally raised successfully to 37 feet and the cross wall bounding the diversion to 33 feet. Just as the team was preparing to celebrate, this work was again submerged in April by a flood which rose 20 feet in twelve hours and carried away two piers of the turbine. The frequent floods caused by torrential rain for the most part of the year proved to be one of the biggest impediments. Even as the site engineers gradually overcame the impact of that flood, more damage was done by another flood later in July.

However, from this point the construction was rapid and uniform. In January 1893, the dam had advanced so far that it became necessary to close several of the vents as they were soon becoming useless. During 1893–94, the absorbing capacity of the lake grew so much that the works never submerged, although the water sometimes passed over that portion on the left flank which had been left low.

Around that time, the work suffered yet another setback in the form of cholera. An epidemic of cholera that broke out at the dam site forced shutting down of operations for about three months that year from March to July as several hundred persons were affected, many of whom died while most others had to be sent back home.

However, work progressed at a brisk pace later during this season. The dam was raised by 47 feet from 68 to 115 feet. As the bed of the tunnel was at 115 feet and the front wall of the dam at 118, it was now possible to turn the water into the plains of Madurai after closing the culvert and allowing the lake to rise.

By July 1895, the water in the lake rose to 110 feet and was passed for the time being through vents previously built. During the remainder of the year, the dam progressed rapidly towards completion, and in October 1895 the dam was formally opened by His Excellency Lord Wondlock, GCIE, GCSI, who laid a stone on the top of the dam in the presence of a distinguished assembly of people to commemorate the occasion. The Mullaperiyar dam has remained active since then.

*

In all such large undertakings, maintaining the health and sanitation of the coolies and workers is a serious endeavour that has a significant effect on the cost of the project. Although hospitals were organized and run with the best possible effort with frequently changing medical staff, it was a difficult task to convince many labourers to accept allopathic treatment. Many deaths were not reported since the patients did not get treatment in the hospital, particularly when an epidemic like cholera struck the worker camps. The very sick among them were removed to the plains by their relatives to die or recover.

Records indicate that in the Periyar dam camp alone there were 76 deaths in 1892, 98 deaths in 1893, 145 deaths in 1894 and 123 deaths during 1895 among a floating population of anywhere between 2,000 and 5,000 workers at that time. While this was the situation at the Periyar camp, the condition of the Thekkady camp was consistently worse. However, the officers suffered less than the workers as they were better fed and clothed, but there was no one who did not suffer ailments during the course of construction.

As bad as the frequent bouts of malarial fever were, the worse killer was cholera, of which there were many sporadic cases. Twice it assumed epidemic proportions. Reporting on one such outbreak, the superintendent of works wrote to the chief engineer on 11 March 1894: 'I have the honour to report that labour has now fallen in consequence of the cholera to such a point that it is impossible to carry on work any longer.

'In an average population of 2,417 (5,000 at the commencement and a few score at the end) there occurred in twenty days, eighty-one cases of which forty-five have ended fatally, not taking into account the deaths which have still to occur amongst the patients still under treatment. This is equivalent to 787 cases and 487 deaths per day in Madras town with a population of 4,50,000. Even these figures are far from representing the real severity of the outbreak, for it is known that many of the coolies were attacked after leaving the camp. Five dead bodies have been reported to me as being found on the roads, one died at Kumili, and there must have been many more cases and probably several deaths.

'For the first nine days of the epidemic, the infected houses were burnt down and their sites disinfected, every hut in the camp was fumigated, medicines were distributed, orders were given to boil all drinking water, drains and latrines were disinfected with quicklime and strenuous exertions were made by cleanliness and any other measures that suggested themselves to stamp out the disease. As these all proved ineffective, it was determined to transport the whole population into a temporary rest camp on the south bank of the river. This was done and immediately resulted in a short lull in the number of cases. The disease soon re-assessed itself however, and after a week it was decided to allow coolies to return to their former camp which had meanwhile been thoroughly sprinkled with solution of corrosive sublimate and afterwards with quicklime. Another lull followed which again proved delusive. There remained nothing to be done except to patrol the lines, with a view to taking each case in good time and to isolate and disinfect, by burning in each case as it occurred. The population continued to dwindle and one line after another to disappear by firing, till there now remain about 200 coolies, the exodus not yet ended, and the camp is merely a patch of blackened ground.'

Besides the deaths caused by diseases, quite a lot of people lost their lives in accidents. A majority of the accidents were connected with the nitro glycerine or detonators and were attributed to the carelessness of the labourers. One of the most common causes for accidents were misfires in the blasting. However, several other accidents including drowning took place at the site during the construction.

The deaths were so many that separate graveyards had to be earmarked for the Indians and Europeans, which can be seen even now at the site of the dam. Following the recent move, the TN state government decided to construct a memorial for Col Pennycuick; the Periyar–Vaigai Farmers' Association members demanded that the memorial should also include all the names of those victims who lost their lives during the construction of the dam and also later during its maintenance.

By the time the project came to its end, the difference between the estimated cost and the actual was so huge that it even drew criticism. However, it was well acknowledged at that time that constructing a dam as huge as this in such extraordinary circumstances of which there had been no precedence in a remote area away from human habitation, expenses were will within limits and the mere feat of achieving the goal of the dam, that is, of diverting the Periyar waters to the Vaigai basin and to Madurai, and the way it has served its purpose for more than 115 years can be called nothing but phenomenal.

*

Within a few years after the Mullaperiyar dam was completed, word spread around that the project was an engineering feat and one of the best in the world at the time. Several eminent civil engineers from across the world heaped praises on Col Pennycuick and some even compared him with Sir Arthur Cotton, a pioneer of large dam works in British India.

At a discussion conducted by the Institution of Civil Engineers, United Kingdom, on the diversion of Periyar in 1897, the president of the organization, J. Wolfe Barry, C.B., had commended Col Pennycuick and his team of engineers for taking the greatest care in making all the engineering calculations, in the tenacity of its purpose and devotion of the staff in undertaking such an effort under difficult conditions.

Lieutenant General Sir Richard Sankey, who had held the position of chief engineer and secretary in the PWD to Madras government between 1878 and 1883, had said that the Periyar project was one of the greatest works of its kind in the world in every direction. He pointed out that the work merited great consideration, especially since Southern India was always threatened by famine. The Lt General also mentioned that the work done by Col Pennycuick was commendable compared to the work and genius of Sir Arthur Cotton, who had worked on several huge projects including the Cauvery River project.

Another eminent civil engineer present at the meeting, George Farren had mentioned that after careful examination, he found the Periyar dam was very nearly what a scientific dam should be. He believed that this dam was the only one that could be called a rational dam in the whole of the British Empire at the time.

The presentation of the project by Col Pennycuick, however, drew considerable criticism too for its deficiency in not harnessing the huge hydroelectric potential of the waters and also in the absence of a tunnel to drain out the reservoir completely in order to take up repair works.

*

To highlight the benefits of a project that has lasted more than eleven decades is silly, at best. If it was not so good or beneficial, the Periyar dam would not be present now and would have given way to younger and sleeker technology long ago.

The project initially envisaged irrigation benefits for 90,000 acres of first crop and another 60,000 acres of second crop. After the main canal and its twelve branches had been aligned and laid out, it was decided to excavate minor distributaries also up to 50 acres limit to expedite the development of irrigation. It was the first time that distribution network was carried up to 50 acres limit at government cost in the Madras Presidency during the period.

The Periyar project also identified the need for a change in the cropping pattern, and the cropping period took some time for the farmers to understand and settle down with. They were accustomed to sowing their first crop in October to take advantage of the northeast monsoon and harvest in January, and resort to a second crop in small areas with attendant risks from February to April. The Periyar catchment being served by the southwest monsoon would yield flows from June onwards and the storage might last up to March the next year which warranted a change in the cropping pattern.

With the advent of irrigation ensuring adequate and timely supply, paddy became the main crop, and the dry crops grown hitherto had to be slowly abandoned. This was also a result of the change in the cropping pattern.

The farmers also faced other problems which retarded the development of irrigation. The people were too poor to undertake land development to convert dry to wet, irrigated land and had to be helped with loans. Since it was not a grazing country and was devoid of trees, plant and animal manure were scant and the farmers had to import manure for which they did not have any resources.

The then government had to import manure and make it available at reasonable costs so that the use of manure for wet cultivation would catch up. They had also agreed to allow a 50 per cent reduction in the water rate for the first three years and another 25 per cent for the next three years in cases where crops were changed from dry to wet.

The year 1896–97 saw the first year of settled water supply to the Madurai region. 50,106 acres of land were irrigated and 7,203 acres of second crop fell under the project command. In the inam and zamindari lands, 1,217 acres of new first crop and 5,225 acres of second crop were also irrigated by the Periyar waters.

During the later years, development showed a steep hike to reach the planned extent under the project, and soon, the Periyar command became one of the most dependable large irrigated extents to be counted in the then Madras Presidency for food production. Over a period of time, ample scope for further enlargement of irrigation benefits from the Periyar waters unfolded and were promptly availed of.

Today, if anyone were to understand the benefits of the Mullaperiyar dam, a brief visit to the farmlands in Madurai and surrounding areas would suffice. The extent of crop that is currently being irrigated using the Periyar waters far exceeds the envisioned extent of land, despite the storage level in the dam being reduced to 136 feet.

As mentioned earlier, the Madurai region was well known for its droughts and famines. The last few generations of farming families in the Madurai belt who get the benefits of the Periyar waters have never known famines during their lifetime. Although the weather is hot and dry as it has been for centuries, their lands are wet and the farmers raise some of the most water consuming crops such as rice and sugarcane despite being in a dry region, thanks to the Mullaperiyar dam.

Another interesting aspect about the construction of the dam is that the present row between the TN and Kerala governments, and the mutual mistrust when it comes to the affairs of the dam was present even 115 years ago at the time of the construction of the dam. This has been documented by Pennycuick in a letter he wrote explaining the escalated costs, citing the non-cooperation from the Travancore government as one of the reasons for not staying within budgets.

In his explanatory note to the chief engineer while submitting the revised estimates, Col Pennycuick states, 'Another cause which has certainly tended towards a general increase of prices is the political condition under which the works have been carried out. This is a subject on which I wish to say as little as possible, because my views are not in accordance with those of the Government of Madras. But it is impossible to avoid all allusion to it. When the estimates were prepared, it was assumed, as a matter of course, that the site of the works and the ground in their neighbourhood would be declared British territory either permanently or for so long as the works were in the course of construction. This was not done, and without going into the details of this portion of the history of the works, it may be said in general terms that, for the first four years of their progress, there was absolutely no machinery for maintaining order or for protection of life and property in the camps, except such irregular and limited criminal jurisdiction as was ceded by the Government of Travancore, but the concessions were so narrow and limited that it was almost worthless, and even at the present time, the conditions of life in the project camps are by no means such as should prevail among a community of some six thousand British subjects. What the financial effect of this state of things has been, it is of course impossible to say with any approach to accuracy; but it is certain that any cause which renders a work attractive tends to reduce prices and any cause which renders it unattractive tends to increase them.'

The TN PWD engineers working at the dam project today would have the same grouse claiming that they do not get any security or protection from the Kerala police who are in charge of the security of the dam. One of the conflicts between the two states is on deciding who will police the dam.

*

Periyar Hydroelectric Power Project

The proposal for a hydro power generating station was mooted by Col Pennycuick even at the time of building the dam. He was aware that power could be produced from the waters that pass through the tunnel passage that he created across the ridge and the 900 feet fall into the ravines before they reached Vairavanar.

The option was referred in 1893 to a committee consisting of Col J. Pennycuick, Prof. George Forbes, Prof. W.C. Unwin and Prof W.C. Roberts Austen. The committee submitted a positive report along with a note on a possibility of power demand arising for production of calcium carbide, aluminium, electric traction on railways, and cotton mills and lighting. In 1897, a pamphlet was also issued by the Government of Madras detailing the report of the committee and calling for tenders for the purchase of the right to develop and utilize power. However, that did not take off due to the poor response to such tenders at that time.

It was later given up as the British government then decided that the power generated might be of use to the locals for their domestic supply, and even if they did need the power, the locals would not have the economic means to purchase it at a rate profitable for the Madras Presidency.

Proposals were again drawn up for power generation using the Periyar waters in 1905, soon after the irrigation project came into operation and were sent to the Government of India. On the advice of the inspector general of irrigation in his note dated 29 March 1906, the Government of India declined to approve the Periyar power project.

The then Inspector General of Irrigation J. Benton in his note opined that the records of flow for the eight years until 1904–05 were too short to take a view whether the primary objective of irrigation would be jeopardized by guaranteeing the flow of hydro power during the years of short rainfall when it was most needed.

Since no user of power would make an offer without a guarantee of constant water supply for electricity generation, which could not be guaranteed in this case, the inspector general considered that the question of generating power be left for a later time. Another proposal was again made by A. Chatterson, the then director of industries, on February 1909 but that proposal also did not materialize for the next three decades.

More concrete investigations into the power scheme were conducted in 1929 and a report was drawn in 1933. Since the power scheme had to be operated only after the irrigation needs would be taken care of, utilizing only whatever discharges were drawn from the lake for irrigation, and since only a maximum of 50 cusecs could be allowed to be drawn in the non-irrigation season, this scheme was also not favoured.

Meanwhile, certain improvements and expansions were underway in the irrigation system in which an intermediate storage was to be created on the Vaigai near Andipatti to store and regulate the flow drawn through the Periyar tunnel, which came in as a complimentary provision for holding on to the tail race discharges of the power scheme even during the non-irrigation season. Due to this, the power drawn could then be increased to a minimum of 300 cusecs.

On the basis of this, another investigation for power project resumed in 1952. Provision was made to increase the discharge capacity of the irrigation tunnel from 1,320 to 1,600 cusecs. Five different sites were explored for the power house and the scheme report finalized choosing the best and most economical site. This scheme was sanctioned and was also included in the First Five Year Plan of the country.

The Periyar Hydro Power Project was inaugurated by the then Chief Minister of Madras K. Kamaraj on 11 February 1955. Actual work on the project commenced soon afterwards.

The main components of the project were improvements to the Periyar Irrigation tunnel of 5,887 feet long to support withdrawal of maximum 1,600 cusecs, construction of a forebay dam across the irrigation channel below the irrigation tunnel, form a forebay of

3.1 m. cubic feet capacity diversion for water, construction of a new power tunnel 4,000 feet long with a surge shaft and pipe tunnel designed for a maximum capacity of 1,600 cusecs, and the erection of four steel penstock pipes, each of maximum 400 cuses discharge capacity, out of which three were proposed in the first stage and one in the second stage.

A power station sufficient to accommodate four generating units of 35 MW each, three in the first phase and the fourth in the second phase, along with the necessary step up transformer and the tower lines, besides a tail race and diversion channel, were all part of the hydroelectric scheme.

After about three years, the first three turbine units were in position by 20 December 1958, and were commissioned one after the other by 25 May 1959. All provisions for the erection of the fourth unit were duly made too. The installation of the fourth unit was however taken up only in 1965, and it was commissioned on 11 September 1965. All the four units, each with a generating capacity of 35 MW totaling up to 140 MW, were ready and operational by September 1965.

The total cost of the Periyar hydroelectric scheme as executed came to ₹356.75 lakhs. The original lease deed of 1886 was later modified to include the hydroelectric project and was rewritten after negotiations in 1970.

*

6

Vox Populi

Travelling in the Dam Country

May 2012

I sent back my part-time driver Kadiresan, a Tamil and a native of Theni district in TN, as soon as I reached Kumily at around 2.00 a.m. Kadiresan was hired to drive me from my home in Coimbatore to Kumily, the scenic hill station situated barely 4 kilometres away from the Periyar Tiger Reserve in Idukki district, Kerala, one of the most popular tourist destinations in this part of the country. Although he had a steady pair of hands behind the wheel and insisted that he accompany me for the three-week trip in Kerala for less than half his wages, I sent him back as I did not want to have anything with me that suggested I could be Tamil.

The last time I visited Kumily was when the Mullaperiyar dam row between TN and Kerala was at its peak and riots broke out almost every morning. This border town situated along the national highway handles most of the interstate traffic between the two states connecting major cities such as Madurai, Dindigul with Kottayam, Ernakulam and the rest of Kerala. During December 2011, when I was here last, vehicle movement between the states was stalled. A large number of police personnel were posted on either side of the border to ensure that no Kerala-registration vehicles went to TN and vice versa.

Overwhelmed at the sporadic outbursts of violence that erupted in Cumbum, Gudalur and other towns in Theni district, I had wanted to see the situation on the other side of the border in Kumily, which was the epicentre of protests in Kerala. My local contact Kumar and I parked our cab about 50 metres away from the police check post on the TN border and walked into Kerala, as dozens of khaki-clad men peered at us suspiciously.

Kumar is from Cumbum and is also a driver. Last year, he bought a second-hand Tata Sumo and made a living by transporting dozens of estate workers from Cumbum, Gudalur, Bodimettu and other parts of Theni to work in the cardamom estates in Idukki district and back. Although he had been shuttling between the two states at least a couple of times every day, two days ago his Sumo was smashed to pieces by a group of protestors near Kumily when he was bringing back the workers.

When we walked past the check post, the police personnel stopped us and wanted to know the purpose of our visit to Kumily. 'I am a journalist and he is with me,' I said. After giving me a hard look, they asked for my identity card before letting me go.

The bus terminus at Kumily is right at the border. We went to one of the tea-shops and asked for tea. Although I am a Malayali by birth and love to speak in my mother tongue at every given opportunity, I spoke only in Tamil. Kumar believed I was a Tamil and had been ranting against the arrogant, selfish Keralites all day. He vented out all his anger in the belief that I was a Tamil and I did not want to shatter him.

Life was as usual on that afternoon in the Kumily bus terminus and there was no semblance of violence but for the huge police presence and a few ransacked banners of shops and broken glass panes. After making a few casual enquiries, we rushed back to our side of the border as I heard that a bus had been burnt in Gudalur became there were a few Keralites in it.

When I reached Kumily this time in May 2012, the first thing I did was to ensure that I spoke only in Malayalam and let go of my Tamil identity for the next three weeks. Besides the registration tag on my car and three full bottles of Bols Brandy distilled and bottled in Puducherry dumped in the trunk of my car, there was nothing to suggest I was born and raised in TN.

Around 10 kilometres away from Kumily is Vallakadavu, a quiet, non descript village located on the Vandiperiyar–Pathanamthitta Road where, according to locals, around 500 families live. To reach here, one has to take the Kumily–Kollam Highway and turn right at Vandiperiyar along a narrow winding road uphill surrounded by sprawling estates on either side. The village is located just ahead of the back gate of the Periyar Tiger Reserve through which the winding road proceeds towards Pathanamthitta.

This village is the first settlement along the Periyar River Valley downstream of the Mullaperiyar dam. If the dam were to break, Vallakadavu and its people would be submerged in 60 feet of water within seven minutes of the collapse.

Thomas Mathew, a stout, balding man in his early fifties, lives along the banks of the Periyar in Vallakadavu with his wife and two of his three daughters. His eldest daughter works as a nurse in New Delhi while the other two are with him. He moved here during the 1970s and currently owns a few acres of land where he grows seasonal crops. For the last ten years, he has been an active campaigner for decommissioning the old dam and building a new one. Having participated in several meetings and protests, Mathew is well-versed with the issues relating to the dam and the fear that haunts their village.

When I met him on a Sunday afternoon at his home, located right on the Vandiperiyar–Pattanamthita Road, Mathew was soft spoken and measured his words. After we exchanged pleasantries, he asked, 'Etha naadu?'

I told him I was from Thalassery and working in Coimbatore for a Chennai-based newspaper. Then we settled down in his living room for a chat. I clarified that having lived for a long time in Chennai, my Malayalam would be bad.

'I like Tamilians, they have loyalty to their state. The politicians and government staff work in the state for the betterment of their people. But here, in Kerala, nobody cares about the public. Even if Vallakadavu gets washed away in the floods and hundreds die after the Mullaperiyar breaks, our people would be happy as they can use the disaster and go begging to foreign countries for relief funds,' he said.

Like most people of his age in Vallakadavu, Mathew is also frustrated with the government and politicians. 'Having been involved in this issue for such a long time, I know of so many dirty games played by our politicians and I can say that nobody is interested in solving the issue. To them, the Mullaperiyar dam is a site meant for disaster tourism,' he said. 'Every year, during winter, they will raise a hue and cry; make the people fools by visiting us and making tall claims about our safety. After a few days, they quietly wither away. Everyone here knows that after these visits, the properties and assets of our politicians and engineers in TN increases mysteriously.'

Over the last three decades, Mathew has seen so many politicians, both from the Left Democratic Front (LDF) and the United Democratic Front (UDF), go to Vallakadavu, climb on a stage, and in pitched voices warn the people there of an impending disaster and how they would be washed away by the Periyar waters soon if the government does not construct a new dam. And then, they go home only to return during the next winter when the dam would be brimming with water again.

But the little children of the village who gather around these makeshift podiums and listen to the tireless rhetoric of the politicians on how their homes would be washed away if and when the dam collapses, go home wondering about their future. On rainy nights, when the sound of gushing waters in the Periyar that flows by their homes disturbs their sleep, these children wake up and cry.

Mathew's eight-year-old daughter Rino Maria is one of them. 'Once she was so scared that I had to take her and the rest of my family to the government school up in the hills so that she felt safe enough to sleep,' Mathew recounts. 'These children have been living in constant fear and uncertainty thanks to the extended media coverage. For the last ten years, Vallakadavu has been in the news constantly, only because if the dam breaks, this village would be submerged under 60 feet of water, and they get to hear and read this everywhere. Even in schools, this is the favourite topic of discussion.'

During an unusually wet season in December 2001, the Mullaperiyar dam filled up faster than usual and soon reached the permitted water level of 152 feet. As the rains continued, fire tenders from the Kerala Fire and Rescue Services were brought from the nearest fire station and stationed right outside the Periyar Tiger Reserve at the end of the village to help rescue villagers in the event of a dam break.

'In the middle of the night, we heard a loud siren from one of the fire tenders. Fearing the worst, we got out of our homes and rushed to higher elevations with our children. But the fire tenders and fire service personnel were ahead of us, tucked away safely inside the school, afraid that the dam had broken,' Mathew recounted with a cynical laugh. 'It turned out that it was an electric short circuit that triggered the siren. The fire service personnel, who were there to save us, rushed out of the spot as soon as they heard the blaring, fearing for their lives. That's the situation here when it rains heavily during winter,' he said.

Over the last several years, this drill has been a common occurrence in Vallakadavu during the winter months. Army personnel have been stationed, fire service personnel and even first aid teams have been deployed to minimize human losses in the event of a disaster. People have also been alerted to move to higher locations when the rains are heavy and there is flooding. But those who live here know only too well that if an actual disaster occurs and over 5 tmc of water flows over their village, nobody would risk their lives to fish them out. They would all be washed away towards the Idukki dam around 40 kilometres away.

Last December, a team of disaster management experts from the state government visited Vallakadavu and apprised the public on the 'do's and don'ts' in the event of a dam collapse and even marked elevations above which they could be safe. Locals say that the government had conducted a simulation exercise of a dam break and arrived at a height of 880 feet above sea level which would be safe for them. Markings have been made across Vallakadavu on lamp posts and compound walls of churches and school buildings located at elevations above 880 feet which are safe destinations to run to should the Mullaperiyar dam actually collapse. 'They have installed a new siren at the government school at a cost of ₹13 lakh. We are told that it would serve as an alarm in the event of a disaster. Hopefully, a short circuit does not damage the equipment next time,' Mathew said.

After a long chat in his living room over the faulty construction of the dam and the huge swindling of money by politicians from both states, Mathew suggested that we go out for a walk across the village. On the other side of the Vandiperiyar–Pathanamthitta Road away from the Periyar River Valley, the elevation increases as we walk along a winding uphill road amid little vegetable gardens, on the one side, and the ABT tea gardens, on the other. On the barks of trees, electric posts and in many of the brick compound walls of homes, 'X' marks painted in white are seen everywhere. They are the heights considered safe for residents—staying beneath this level is dangerous for the residents—to rush to as soon as possible to avoid getting washed away.

About 2 kilometres uphill on the narrow winding road, Mathew asked me to stop and pointed towards a distant grey line in the midst of rolling hills deep inside the Periyar Tiger Reserve. 'Do you see that grey wall? That is the Mullaperiyar dam, and on clear days during winter, we can see it filled with water,' he said.

Looking down, I noticed that we were on the top of a little hill, and Mathew's home was way down below at the bottom of the valley. If that dam broke, I was now certain that Mathew's home would be under several metres of water.

This viewpoint in Vallakadavu is the closest point from which one could legally see the Mullaperiyar dam. Visitors are not permitted by the Kerala forest department to venture anywhere into the Periyar Tiger Reserve except the front entrance meant for tourists through Kumily, where a boat ride would ensure a similar distant view of the 119-year-old dam. Actual field visits are permitted only on special occasions, such as visits by politicians or engineers involved with the maintenance work.

After walking down the hilly road, we quietly sneaked into the tiger reserve through a crack in the fence known only to locals. 'We know most of the forest guards and they don't mind us getting in as long as we remain on the border,' he said. The tar road towards Pathanamthitta branches into a mud road that makes a hairpin bend into the jungle. A security check post with a forest guard sits at the entrance to it. 'This is my distant relative. He wanted to see the forest and I brought him,' Mathew told the guard, who smiled and waved back. The guard warned us not enter further inside the forests. 'I will get into trouble if someone sees you,' he said.

We walked back towards the back entrance of the reserve along the course of the river that has now reduced to a mere trickle. Mathew pointed at the little stream and said that all the water that flowed through the stream was seepage from the dam. 'If there is so much seepage even during summer, you can imagine the condition of the dam,' he said.

As we got out of the tiger reserve, the first home outside the giant gates, probably the one that would be washed away first, belongs to a Tamil family settled in Idukki district for generations. Not just the first home but quite a few homes bordering the village belonged to Tamils settled in Kerala for generations. We walked into one of the homes and Mathew asked me to find out from them about what they felt about Tamil politicians who were fighting against Kerala.

Even as I tried to talk to them in Tamil to find out how they felt living right under the nose of the dam, the women who were present in the home replied in Malayalam and asked me to leave. 'We don't want to say anything, come back later,' they said. These women were better than me at trying to be a Malayali. Like so many other poor migrants across India, they too know that no native cares for their opinion.

On the way back, I met Michael Joseph, a self-styled sanyasi who was a chartered accountant in New Delhi before he gave up urban life to settle down at Vallakadavu. He lives a few homes away from Thomas Mathew's place, along the main road. Lean and long-bearded like an activist, Michael lives in a one-room apartment and moved here three years ago as part of Infarm (a study of emerging farmers' movements in Kerala) to take up religious as well as social activities in Vallakadavu, and has been a vociferous campaigner for a new dam. Among other things Michael organizes prayer meetings among local Christians and is an influential person in the village.

Like most others in the village, Michael too blamed the 'impotent and selfish' politicians of Kerala for not resolving the contract for so many years. 'We don't trust the politicians, we don't trust the bishop, and we don't trust the activists or anyone. Nobody is interested in protecting the interests of Kerala,' he said. "But Tamilians are not like that; they are very selfish and take care of their interests irrespective of whether the truth is on their side or not,' he said.

As he continued to vent out his hatred for Tamils, I started getting restless. For someone who has lived in and loved the land where he was born, listening to others hate it made me uncomfortable. At one point, he even went to the extent of pointing out that Chinese and Pakistanis would be better neighbours than Tamils. 'Can you imagine, they have been drinking our water and getting prosperous for a hundred years and now, when we want to build a dam for our own safety, they would rather have us killed? What kind of people are these?' he asked.

Michael pointed out that Malayalis had at least ₹15,000 crore worth investments in TN and would do nothing to jeopardize that money. 'Take, for instance, MRF or Malayala Manorama or even the Catholic church that have made huge investments in TN. Why would they ever want to hurt their investment by going against Tamils? Nobody would come in support of their own state when it comes to the Mullaperiyar issue,' Michael said.

According to him, almost anyone who had even mildly spoken in favour of TN in a public forum, including respected Justice K.T. Thomas of the SC, was a Tamil stooge and had taken huge sums of money.

*

The drive between Vallakadavu and the Idukki dam, along a stretch of around 40 kilometres through narrow winding roads that pass through steep hills and valleys, is breathtaking. On one side are thick, dense jungles with tall, ancient trees and evergreen foliage interspersed by colourful flowering plants and little front yard gardens while, on the other, is the steep, deceiving valley of the Periyar. On rainy days, the river unleashes its full fury with a deafening, constant roar.

Along this beautiful stretch of road lie the villages of Vandiperiyar, Keerikara, Mlamala, Chappathu, Parappu, Uppukara and Ayyapankoil. The proponents for constructing a new dam at the Mullaperiyar strongly believe that if the existing dam breaks, flood waters would gush from the Periyar dam towards the Idukki dam, located more than 40 kilometres downstream the Periyar, wiping away all the villages on its way. Assuming that the Idukki dam is also full at the time of the collapse of the Mullaperiyar, the extra 5 tmc water would not be withstood by the more modern Idukki arch dam, resulting in the release of 70 tmc of water that could wash a significant portion of the state of Kerala including some top towns such as Kochi.

While a multi-dam failure could be considered as stretching the probability theory a little too far, the fact that the villages between Vallakadavu and Idukki dam would be wiped away if the Mullaperiyar dam gives way is a plausible scenario.

While the protests against TN and its stand on the Mullaperiyar dam row have always garnered the full support of people from all these villages, the quiet, nondescript village of Chappathu, about 25 kilometres away from the Mullaperiyar dam, has been the epicentre of these protests.

For the last six years, a relay protest has been continuing in a permanent shelter put up by the members of the Mullaperiyar Samara Samiti along the banks of the Periyar in Chappathu village, where at least a couple of men or women come every day and observe a token protest from 9.00 a.m. to 5.00 p.m. This protest has been going on since 2006 and is symbolic of the persistence of its organizers in never giving up their demand for a new dam.

When I went to the protest pandal at around 4.45 p.m. some time in mid-May 2012, two to three men who had sat in protest through the day were getting ready to call off the fast only to continue the next day. A police inspector, a jeep and a few constables have also been permanently posted at the pandal and guard the venue round the clock. The river bed adjacent to the protest site was almost dry barring a few thin streams piercing through the cracks of huge, curvy boulders that resembled giant eggs and lay underneath a concrete overbridge road. On the other side of the bridge, another narrow winding road led to the interior parts of Chappathu.

Chairman of the Mullaperiyar Samara Samiti, Father Joy Nirappel, who is also the Uppukara head of the Roman Catholic Diocese, had accompanied me to the permanent protest shelter at Chappathu in my car that evening. Fr Joy is a short, stout man with a greying goatee beard and a deep, mellow voice, which he has learnt to modulate perfectly over years of public speaking. He was one of those few persons who were friendly and willing to give an interview, unlike many of the leaders of protest groups connected to the Mullaperiyar dam. He seemed to be in command of the village and its people, until we reached close to the venue.

'I don't know what these people might say if they see both of us travelling in a car with a Tamil Nadu registration,' he said.

'Why?' I asked, 'Is there a problem?'

'No, no…nothing like that. It's just that so many rumours make the rounds around here that people are so confused and could be easily misled,' he said, and looked at me reassuringly.

As soon as we reached the venue, Fr Joy and I got out of the car and went to the protestors and greeted them. 'This is Pradeep Damodaran, a Malayali journalist from Thalassery. He has come to interview me about the Mullaperiyar dam and our protests,' he said, introducing me to the handful of people seated at the pandal. I shook hands with them and clarified that although my native place was Thalassery, I was working in Coimbatore and had come from Coimbatore to do the book. After a few minutes of uncomfortable silence, I decided to leave the place after clicking a few pictures of the historic protest site, and Fr Joy looked relieved.

The Mullaperiyar Samara Samiti was formed in the year 2006 at a meeting held in Karol Plaza, Upputhara on 3 March 2006. At that time, there was another organization fighting for the rights of people living downstream the Periyar, called the Periyar Valley Protection Movement headed by late A.T. Thomas. The Samara Samiti was mainly formed following the 2006 verdict by the Supreme Court of India granting TN rights to raise the water level in Mullaperiyar dam to 142 feet.

Public resentment over the SC verdict in response to two petitions filed by the Mullaperiyar Environment Protection Forum and Subramanian Swamy, allowing TN to raise the water level to 142 feet, that was delivered on 27 February 2006 became the inspiring force behind this relay strike that has reached historic proportions and has been continuing for over six years now.

Just a few days after the verdict, members of the Periyar Valley Protection Movement, the organization that was spearheading the campaign for decommissioning the Mullaperiyar dam, headed by A.T. Thomas, convened a meeting at Karol Plaza in Upputhara on 3 March 2006. 'Mullaperiyar Samara Samiti was formed on that day to carry forward our fight for a new dam and I was appointed as chairman of this committee,' Fr Joy said.

The protests were organized in three phases. 'The first phase was one of aggression. It was the need of the hour at the time. Our fight for the safety of the people living here was going out of our hands and we wanted to bring it to the notice of the state government. From 3 March, he held protest marches, formed human chains on the road, blocked roads, blocked national highways, sent hundreds of letters to the prime minister, and so many other things,' Fr Joy said.

More than 10,000 people marched to the venue of protests every day, from Vallakadavu, Vandiperiyar and the surrounding areas, to participate in the protest meets that gathered momentum with every passing day. The Samiti members also organized a march to the state secretariat besides conducting state-wide rail roko exercises.

'The protests were a great success for the Mullaperiyar Samara

Samiti. All the print and visual media were here and covered the protests round the clock. Six to seven MLAs, and several MPs and ministers visited the protests and expressed their support. Eventually, the state government, that was desperate to silence the protestors, had passed the amendment to the Irrigation Act, restricting the water level to 136 feet and prevented TN from implementing the SC verdict. The matter went back to the courts, and following that success, our protests moved into its second phase of a relay protest,' the Uppukara head priest said. As he spoke of the protests and their success, Fr Joy's voice rose to the level of public speaking, and he seemed to enjoy listening to it as much as I did.

A permanent venue for the relay protest was chosen at Chappathu on the banks of the Periyar. A shelter was erected temporarily using poles and tent material and was inaugurated on 25 December 2006. With the setting up of the permanent shelter, the Mullaperiyar dam became a part of the psyche of the people who live in this village and the surrounding areas, constantly reminding the public of the government's lack of concern for their plight and their need to unite and fight against TN.

For six long years, the permanent shelter has endured sun and rain, LDF and UDF, and several other forces of nature, making a statement on the resolve of the locals here and their campaign for a new dam. While sporadic incidents of violence have been repeatedly recorded as a fallout of these protests in the form of attacks against TN registration vehicles, etc., the protest, as such, has been an excellent exercise in expressing the democratic right of an individual or group of people to protest against the state and the central government.

There is probably no other venue in the world where so many different forms of protest have taken place, according to Fr Nirappel. 'We have conducted protests by lying inside a coffin, by holding our breath and drowning inside water, human chains, our members have stood on one leg and protested, some of them have stood upside down for hours together, etc. We have boycotted public events, boycotted football matches and so many other forms of protests. I can boldly say that there is probably no other venue in the world where so many different protests have taken place over the last six years of our existence,' he said.

A large number of Tamils living in Idukki district have also been active participants in the protests and have been fiercely campaigning for a new dam for decades. Unlike any other part of Kerala, Idukki district is probably the only district in the state with a large Tamil population. Tamil families have been living here for several decades now and many of the settlements are even older than the Keralites who began moving up to the hills only during the mid 1900s.

While the region that is now called the Periyar Tiger Reserve where the Mullaperiyar dam is located, and its surrounding areas were thick, uninhabited jungles at the time of construction of the dam, around the same period of time several cardamom and other plantations started coming up on the hills, that required a large number of casual labourers. The nearest and most accessible source of labour was from Madurai and Theni districts from where thousands of families of Tamilians moved to the upper slopes of the hills and lived and worked in the plantations and tea estates. For generations, these families have been living here and are as much Keralites as they are Tamil.

Until the 1950s, not many Malayalis were living in Idukki district, mainly due to the more prosperous coastal belt. 'When famines and unemployment levels rose in the plains, quite a few families moved up the hills to make a new life for themselves. Upputhara was one of the first few villages that came up on the banks of the Periyar, where I know people who have lived for more than seventy-five years,' Fr Nirappel said.

But large scale migration of Malayalis took place only post 1950s, when the politicians at the time felt that there was a big imbalance between the ratio of Tamils to Malayalis in Idukki district. During that period, many people who were not very successful in the plains were encouraged to move to the Periyar Valley in Idukki district and were given up to five acres of land on a cheap rate to cultivate crops and be self-sufficient. The population in many of the villages bordering Periyar River Valley swelled during this period. The church plays an active role in community activities and has a good rapport among residents, as is evident in the role it has played in the long life of Mullaperiyar Samara Samiti.

Even as the second and longest phase of the protest, that is, the relay protests, had been going on, certain events that took place in November 2011 once again altered the phase of the protests.

Between 1 November and 20 November 2011, minor tremors were experienced in several parts of Idukki district; many of which were not even recorded by geologists. But the residents living along the Periyar Valley were terrified at the prospect of a larger quake at a time when the dam was almost filled to its permitted capacity. Soon, word spread around and the permanent protest shelter built by the Mullaperiyar Samara Samiti at Chappathu once again began to gather crowds, most of whom were angry.

'Our third phase of protests began on 27 November 2011, with the announcement of an indefinite strike as a result of public outcry following the recent earthquakes,' Fr Joy said. 'This phase of the protests was more potent and drew more crowds than ever before. Keralites from Kasargod to Trivandrum participated and joined in the fight for safeguarding our lives.'

As with every other earlier occasion when the protests gathered momentum, an array of politicians ranging from local MLAs to MPs to state and central ministers visited Chappathu and expressed their solidarity with the cause of the protestors. Every day, thousands of people participated in hunger strikes, road rokos and other forms of protest. By early December that year, protests began to spill over onto the streets as buses plying to TN were scribbled with abusive, vulgar messages and sporadic incidents of attacks on vehicles with TN registration were being reported in the interior parts of Idukki district.

Although coordinators of the Mullaperiyar Samara Samiti officially abhorred violence and claimed that they practised only protests following Gandhian principles, fringe elements within the group did form small groups and organized protests. Dozens of vehicles bound for TN were stoned and attacked with wooden logs. Soon, the attacks were retaliated from across the border and the issue escalated into a full-fledged riot in a matter of days.

According to Fr Joy, the third phase of protests lasted fifty-three days and was a grand success. 'The support we received from Keralites not just in the state but also from the Middle East and even the United States was overwhelming. The protests peaked in December and reached a point when the attention of the entire country was diverted to the Mullaperiyar dam and the safety of thousands of people who live downstream. When things reached a climax, we decided to pull down the ferocity of our protests from 17 January 2012,' he said.

Some of the main reasons that were cited for drumming down the tempo were that the empowered committee was expected to present their report during early February 2012 and the matter would go to the courts again for trial. 'So we decided to wait and watch. The protests returned to the longer, less intensive phase, and we are still continuing with our relay protests,' Fr Joy said.

At the time of visiting Chappathu, the venue had seen 1,969 days of protests, and the Samiti members were planning for bigger events when the protest completed 2,000 days.

The Mullaperiyar Samara Samiti has laid down clear conditions to the governments of Kerala and TN for them to give up their indefinite strike. They include:

(i) Inviting an international agency to test the safety of the dam and allow them to conduct their test independently, without any intervention from the CWC or the TN engineers. The Samiti members claimed that they have lost faith in the CWC and that the public will not believe any tests conducted by any Indian agency. If the international agency states that the dam is safe, then the Samiti is willing to let go of its demand for a new dam;

(ii) Completely rewriting the lease deed of 1886, signed between the Maharaja of Travancore and the Madras Presidency, and change the term of lease to a much lesser period. The Samiti pointed out that the lease deed had to be completely rewritten as even if a new dam was constructed using the same lease deed, again similar protests would erupt a hundred years later as the term of lease is for 1,000 years;

(iii) Ensure that the state of Kerala is adequately compensated for the water supplied to TN and the power that is generated at the Periyar hydroelectric project. The present financial terms of agreement are unfair and need to be revised; and

(iv) Attempt an out-of-court settlement between Kerala and TN in resolving the issue amicably by attempting to solve the earlier mentioned issues.

Until a permanent solution is arrived over the issue, the Samara Samiti has also issued a set of safety measures to be taken up by the Kerala government to ensure that loss of life is minimized in the event of a disaster.

'We have asked for installations of sirens in every village to give an alarm to the public on when they need to run out, setting up a mobile phone alert facility where messages could be sent to all cell phone users, installation of solar streetlights, permanently stationing a fire service team and mobile ambulance team, constructing permanent shelters at higher elevations to move injured and sick persons along with women and children, provide disaster management training to all the residents and counselling for our children to deal with the fear psychosis, among other demands,' said Fr Joy, as he read out from a long list, neatly typed and filed in one of the dozens of files neatly arranged on his table.

While a few of the measures have been taken up, such as installation of sirens at Vallakadavu, most other demands of the Samiti continue to remain only on paper, and the protestors seem to be clear that the protests would continue as long as their demands are not met and people can feel safe living in their neighbourhood.

'As of now, the protests are going through a smooth phase. We will enhance its vigour as and when we feel that the government is failing us, and the Mullaperiyar dam issue will once again became national headlines,' Fr Joy said.

As dusk set in, the protestors along with Fr Joy left the venue and moved in different directions. A few chairs remained strewn around inside the pandal, while the lone police inspector and constables stood around the police jeep and engaged in chit-chat.

*

Senthilvelan from Vandiperiyar cannot recall when his family moved from TN to Kerala. The tourist taxi driver, who currently lives and works in the scenic hill station of Munnar, was born and raised on the banks of the Periyar about 5 kilometres away from Vallakadavu, and is only too familiar with the 'hate Tamil' campaign that has been brewing in his backyard for the last three decades.

'My father was also born in Vandiperiyar but his parents came from Theni district,' he says. 'Although we do have some relatives back in Theni, I rarely go and visit them. My life is here and I do not want to leave Munnar,' he says.

Senthil's grandparents moved to Idukki district to work in the tea and cardamom plantations, where they were provided homes and given free education by the plantation owners. Over the years, the plantations kept changing hands and many of the former estate workers moved on to other businesses, such as small hotels, textile shops and other establishments. His father, too, worked as a daily wage labourer in several plantations in and around Kumily until he retired a few years ago. 'I did not want to do that work and so, I learnt driving and moved to Munnar,' he says. 'Here, I get regular tourist bookings and am busy for twenty days in a month during the busy season. Even if I am busy for ten days in a month, I would have made enough money to live. I go back home to meet my parents once a month, otherwise I am very happy here.'

I met Senthil on my way to Munnar from Kumily at a roadside eatery. He had come back home to visit his parents and sister, who has been married off to a distant relative in Peermedu and was back for the summer vacation.

For the first time in over a week, I spoke in Tamil, a language that I had grown up with and am most comfortable in. Hearing my accent, Senthil too became a little comfortable with me and started chatting. When I asked him about the Mullaperiyar issue and if they felt safe living in Kerala where there is so much hatred against Tamils, he said that the state was his home. 'I was born and raised here. Nobody will attack me or my car in Vandiperiyar,' he said.

Amid the relay protests in Chappathu and the hate campaign against Tamils, thousands of Tamilians continue to live happily in Devikulam, Peermedu, Vandiperiyar and many other parts of Idukki district. The tea-shops and roadside eateries all serve authentic Tamil food and the name-boards are also written in Tamil. The locals here claim that none of their shops or businesses was attacked. 'They were only targeting Tamilians from Tamil Nadu, not us,' Senthil said.

Most Tamils living in Idukki district agree with the stand taken by the Kerala government on the Mullaperiyar. They are baffled at the rhetoric made by politicians in TN when they claim to be fighting the Tamil cause in Mullaperiyar. 'We are Tamilians too. If the Mullaperiyar dam were to break, an equal number of Tamil lives also will be lost. Besides, Kerala is anyway willing to give you water,' he said.

The word 'you' shocked me. Ever since I've been covering the Mullaperiyar dam row, I have totally got my identity confused. When I am in TN, the activists and protests brand me as a Keralite and say 'you Malayalis', and when I am in Kerala speaking on the Mullaperiyar issue, the Malayalis say 'you Tamils'.

Here, in Peermedu, a Tamil taxi driver born and raised in Kerala addressed me, a Malayali born and raised in TN, as 'you Tamilians'. For once, I thought, someone was making sense.

*

Every morning as I checked out of my moderately priced, single bed, non air-conditioned room at a private lodge that I had booked for a week in advance at Kumily, the manager would reconfirm with me if I had shortened my stay. 'We are getting a lot of queries. I just wanted to check with you,' he would say, with a polite smile.

During my stay, thousands of tourists thronged Kumily every day. The Periyar Tiger Reserve in Thekkady was so crowded every other day that tourists used to wait in the queue for a boat ride as early as 6.00 a.m. Hotels were booked months in advance and any late arriving tourists without a booking were ripped off royally.

While domestic tourism reaches its peak during the summer months, there is a steady flow of local and international tourists in Kumily throughout the year. The entire town is filled with motels, lodges, luxury spas and high-end bed and breakfast resorts all the way up to the Periyar Tiger Reserve in Thekkady located about 4 kilometres away.

For most people in this town, tourism is the only business. One can find young men luring tourists by selling anything ranging from a bunch of cardamom seeds to special marijuana, which they claim has been grown deep inside the Western Ghats and known to possess exotic properties. Exotic massage parlours and Ayurveda therapy centres can be found in every nook and corner of this town catering to the wealthy, affluent tourists.

Local residents are mostly confined to the interior areas, and commute to Kumily to work in the tourism industry. Hence, even as the protests became fierce and gathered momentum during December 2011 across Idukki district, the situation in Kumily that is located on the other side of the Periyar dam, was not as tense.

Just a decade ago, a Kumily-based environmental organization, Mullaperiyar Environment Protection Forum (MPEF), spearheaded Kerala's protests against TN's demand for raising the water level from 136 to 142 feet and fought the case at the highest level in the SC. During that period, neither the Mullaperiyar Samara Samiti nor any other organization was active and all protests and demonstrations were focussed in and around Kumily.

Chairman of the Mullaperiyar Environment Protection Forum Mr Joseph Karoor lives in Kumily, and has been fighting to protect the tiger reserve and the fragile ecosystem inside the reserve forests for the last several decades. He is also one of the few opponents of a new dam.

'We are in no way connected with the Samara Samiti and other organizations. Our primary interest is in protecting the environment and prevent it from degrading,' Mr Karoor said. 'Our primary contention is that the water level in the Mullaperiyar Dam should not be raised beyond 136 feet, and Tamil Nadu should ensure that the dam is safe by doing proper maintenance,' he said.

Karoor and his associates at MPEF had filed a case at the Kerala High Court, some time in 2000, demanding that the water level in the Mullaperiyar dam be retained at 136 feet, as a counter to the petition filed by Subramaniam Swamy demanding that the water level be raised as all the strengthening measures advised by the CWC were completed by TN.

After both the cases were referred to the SC, members of MPEF fought a legal battle for six long years and eventually lost the case to TN, as the courts declared that the water level could be raised to 142 feet.

'Our main contention is that the water level should not be raised. Over the last thirty years, so many plants, animals and insects have been used to the shrunk reservoir space due to the reduction of water level in the dam. If we suddenly raise it, several acres of forest would be submerged in the water, killing a variety of species,' the MPEF chairman said.

On the question of the safety of the dam, the MPEF chairman does not want to take a stand. 'Nobody can predict when a dam or any building would collapse. I might say that this dam will break tomorrow but the dam might go on to live for another hundred years, and at the same time, we might give this structure a life of another fifty years and the dam might break tomorrow,' he points out. 'But we are definitely against the construction of a new dam, and we will oppose it if the Kerala government decides to proceed with their plans,' says Mr Karoor.

According to him, the proposed site for the new dam is located in the core area of the tiger reserve, and constructing a huge dam in the middle of a reserve would further be catastrophic for the wildlife living there. 'Firstly, it is extremely difficult to get environmental clearance for a new dam, and even if they did get it, it would be difficult to get public support as constructing a new dam could do huge damage to the tiger reserve,' Mr Karoor said. 'Dams are always good for governments, politicians and engineers as everybody makes money out of it. Thousands of crores of rupees would be pumped into the project and several people will make crores before the dam is built. The Idukki dam took thirty years to complete; this will take at least half of that time and the Periyar Tiger Reserve would be finished.'

MPEF members want the TN and Kerala governments to refrain from increasing the water level to 136 feet, or even lower, depending on the needs of TN, and ensure that a reasonable amount of water is allowed to flow into Vallakadavu and surrounding areas as they claim that there is water scarcity in Vallakadavu, Vandiperiyar, etc. during summer months.

'Spillways could be strengthened, dam structure should be properly maintained, and Tamil Nadu should reduce its dependence on this water by changing their crop pattern. This is the only possible solution to the problem. It is basically a case of misunderstanding,' said Mr Karoor.

Currently, MPEF has only thirty active members, most of whom are situated in and around Kumily. 'But this is a very sensitive issue and when we raise it, huge crowds would gather as we are basically fighting to protect our environment, which is the source of livelihood for many in Kumily and a major revenue earner for the state,' the chairman said.

While there is widespread resentment in Kumily, as in the rest of Idukki district, on the unfairness of the 1886 lease deed and the power loss to Kerala due to the Periyar hydroelectric project, the main contention of the Kerala government and the other protest organizations, that is, the demand for a new dam, does not have many takers in this tourist town.

A Stroll among the Rice Fields of the Periyar Basin

Five months after I was smoked out of my hotel room in Cumbum by a group of rioters who suspected me to be a journalist working for a Kerala-based newspaper and threatened to attack, I returned to the scenic Cumbum valley during May 2012 to find out what drove these otherwise normal, passionate people to indulge in violence and anarchy on such a scale.

The crowded and dusty town centre hardly resemble the curfew prone area that was flooded with policemen when the riots were at its peak. Normalcy returned to the border town, although certain scars remained even after such a long time to haunt the memories of the victims of the December 2011 riots. Broken windowpanes of shops owned by Malayalis, torn down banners of Muthoot Finance office and other shops remained unfixed. The non-vegetarian hotel run by a Keralite that was famous for its beef curry and cuppa has now been converted into a vegetarian South Indian hotel after the management changed hands. Many Keralites moved their businesses and homes to other areas fearing another backlash. Otherwise, KL-registration vehicles continue to zip past in the winding Cumbum–Kumily highway, and people go about their business as usual. And in street corners and tea-shops, conversations returned to the latest films and cricket matches and woes of farmers.

I met Ramakrishnan, a successful farmer from Cumbum and local leader of the MDMK, the party that was most active during the riots, outside a public hall near the government hospital. Hailing from a family of farmers for three generations, Ramakrishnan is an active member of the Cumbum unit of the Periyar–Vaigai Basin Farmers' Association; he was an active participant in the massive protests that broke out last year.

At first look, he did not seem the kind of guy that went to the streets to protest. The prosperity of his ancestors as well as his own was evident in the ample flesh that tagged to his body and the neat white shirt and dhoti that he wore. Ramakrishnan does not recall a time when the Cumbum valley was impoverished.

Ever since his family moved to Cumbum way back in the 1930s, they have been enjoying access to the Periyar waters, and he attributes their success and prosperity to the steady supply of water for ten months in a year. To him as well as thousands of other farmers in Theni district, the Periyar waters is a source of livelihood and the Mullaperiyar dam is an engineering feat that has transformed life in this region.

'We have not known any famine. My grandfather as well as my father have never told us about any famine affecting this region. In fact, one of the reasons for my ancestors to move to Cumbum could have been the guaranteed supply of water,' he said. 'But our leaders tell us that this region was prone to violence and bootlegging and waylaying before the dam was built, as people did not have a steady occupation due to lack of water.'

During most years over the past century, water from the Mullaperiyar dam was released on 1 June and was continually supplied until 31 March. April and May are the dry months where there is no work in the farmlands here. Otherwise, farmers from this region get two harvests, and rice paddy and sugarcane are the most popular crops here due to the abundant supply of water. Being the village closest to the dam site, Cumbum, Gudalur and surrounding areas receive water even if the water level in the Mullaperiyar dam is as low as 116 feet.

As a result of this guaranteed supply of water, at least five lakh families in Theni district depend on agriculture. Farmers who own lands employ different sets of workers through the year for sowing, reaping, harvesting, preparing fertilizers and many other farm related activities; and according to locals, if the farming activity comes to a standstill on a particular year, many people are forced to migrate as there is hardly any job or money rotation in the villages.

Ramakrishnan said that he provided employment to at least a dozen families every year. 'When I have to sow seeds, I approach a contractor and get around thirty–forty women to work in the fields. After that season is over, I need men to maintain the crop and ensure it is ready for harvesting. Then, we need brokers to take the seeds to the market, mill works to polish it and so forth. I can safely say that at least 80 per cent of employment in Cumbum and surrounding areas are generated through agriculture. If not for the farmlands, I do not know what our people would be doing. Thanks to the Periyar waters, we are able to engage in a respectable occupation and lead peaceful lives,' he said.

When the respectable life that he led pursuing the noble occupation of producing foodgrains was threatened, even the suave, white-shirt-white-dhoti-clad Ramakrishnan took to the streets and protested against the move by Kerala during December 2011.

'You would not believe,' he told me excitedly, 'every day, I, too, walked, along with the protestors against Kerala's proposal for a new dam, from Cumbum all the way to Kumily. We walked all day and reached there by evening, then staged protests and returned at night, only to go back the next day. Lakhs of people gathered together and participated in the march as the entire country watched. I think we have finally made it clear to Kerala as to how dear the Mullaperiyar waters are to us and what would happen if we are denied access to it. If we had not reacted forcefully, our future generations would blame us as the ones who had lost the Periyar waters and pushed them back to an illegal way of life. Take my case—if I have to give up farming, I would have to either think of sand smuggling or some other illegal activity. I cannot get a fresh degree and go for a white-collar job at this age and I have to feed my family; what would I do?'

Unlike other parts of TN, Theni district does not have any major industries or other sources of employment. The few mills that operated out of here had already closed down, and most farm hands now depended on the central government's National Rural Employment Guarantee Scheme (NREGA) to make a living when there was no work in the farms.

Whenever the farming activity was erratic due to lack of rainfall, as was the situation a few years ago, many farm workers moved to other areas to work in mills, industrial units and in the construction industry. 'We do not get as many farm hands as in the past. The sector is already reeling under a crisis and the Kerala government's proposal to build a new dam could be the last nail on the coffin,' Ramakrishnan said.

But that was not his primary concern when I met him. Over the last several months, the man has been desperately trying to find a seat for his son in a reputed engineering college. 'I have been trying through all my contacts but have not received any positive response. The higher secondary school leaving exam results will be out tomorrow. I want to ensure that he gets admitted in a good college,' he said.

I asked him if he did not want his son to do farming, like his forefathers and himself. 'No way, it is a really hard job. I own ten acres of land and work hard on it for ten months in a year to earn only a few lakhs of rupees. And that too is becoming unpredictable with every passing year. I want him to do engineering and probably go abroad and make a good life for himself,' he said.

Before leaving Cumbum, I tried to meet Kumar and to clarify one last time. He attended my phone call and remembered me too.

'How are you, sir?' he asked.

I told him that I was fine and wanted to meet him. He agreed to meet me later that evening at around six. He also told me that the police had still not registered a case for his damaged Sumo and that he had decided to move on.

Later that evening, I waited to meet Kumar near the government hospital where I first met him, at 6.00 p.m., as he wanted me to. Kumar never came and when I tried calling him, he did not pick up my calls.

*

The official bungalow of the Theni superintendent of police (SP) is tucked away behind the sprawling office of the SP, and can be reached by driving along a narrow, gravel road overlooking vast stretches of vacant land a few kilometres away from Theni town.

On my way back from Cumbum, I visited SP Pravin Kumar Abhinapu on a warm Saturday afternoon at his bungalow, mainly to see him in person and to thank him for helping me get out of the situation without getting hurt, and also for doing a little probe on how my Facebook photograph had reached the streets of Cumbum and Palarpatti.

While I had spoken to Pravin Kumar over the phone several

times while covering the riots, I never really had the opportunity to meet him or another police officer who was at the scene, because covering the riots itself took all my time.

SP Pravin Kumar is unlike any other police officer I have met so far. He looked more like a software engineer—young, energetic and extremely chilled out. He is probably the coolest IPS officer in TN cadre and that probably helped in bringing down the temperature faster than expected in this steamy, arid district.

After we shook hands and he settled down behind the large desk at his home office, dressed in a polo t-shirt and shorts, I thanked him for helping me out.

'Boss, trust me, you were the least of my worries at that time,' he said. 'We did probe a little bit into how you were targeted by miscreants but did not get any major leads,' he said.

The police, however, had tracked down the youth who showed my photograph at Palarpatti and asked if I was a Keralite trying to spread wrong information. 'The boy told us that he was returning home from Gudalur the previous night and got down from a bus at Cumbum bus stand when a group of people were distributing photocopies of your picture and asking locals not to entertain you in any way. He just picked up a copy and came home. To his surprise, you landed up in his village on the following morning,' Pravin said.

Unfortunately, the police could not make any more headway into the investigation, and all we could do five months later about it was have a nice laugh over the incident.

SP Pravin Kumar said that he had received intelligence reports even as early as late November that trouble was brewing on the other side of the border. 'As the protests gathered momentum in Kerala, we got information that all interstate buses that were returning from Kerala contained slogans and abuses against Tamil people and Jayalalithaa that were scribbled on the bodies of TN-registration vehicles. In order to prevent the situation from deteriorating, I even put a ten-member team with a bucket of water and cloth at the interstate border to stop all buses and clean up the abusive scribbling before the buses came into Tamil Nadu. But soon, things went beyond our control,' he said.

Even during the first week of December, that is, when I visited Cumbum, the violence was only sporadic and unorganized. 'We still thought we could manage the situation. Although there were a few attacks on vehicles and some incidents of arson and looting in Cumbum and Gudalur, the situation was not grave, and the Idukki superintendent of police and I even had a meeting to discuss and take control of the situation. Looking back, it sounds as the silliest thing that we could have done as all hell broke loose by the next week,' Kumar said.

While several small-time politicians and leaders of fringe outfits held regular meetings and stirred the feelings of the public, the final trigger that blew the situation out of control was the decision by the Kerala state government to reduce the water level in the Mullaperiyar dam to 120 feet.

'After that news reached Cumbum, Bodi and other areas, thousands of people gathered every day and marched to the Kumily border carrying broomsticks, logs, etc. These events were so sporadic and organized in such a short time that despite introducing a number of intelligence officials, we could not gather any information about the protests beforehand,' he said.

'Every morning, I would get to know of the plans of the protestors only by around 8.30 a.m., and we would immediately gather into firefighting mode. By the time we mustered the strength necessary to silence the protestors, it would be three or four in the evening, and on the next morning, the situation would go back to square one and the protestors would start a fresh march,' SP Pravin Kumar said.

One of the reasons for the intelligence network to not have been so effective, according to me, is because most lower level policemen were with the protestors and were of the opinion that the protestors were right about what they did, especially the police personnel who came from down South and were aware of the issue.

When the riots were in full swing and a one-day strike was being observed by the auto drivers in Uthamapalayam, I had stopped by at an auto stand to check out what was happening as a crowd had gathered around a TV camera crew. A vernacular TV channel reporter was interviewing the leader of the autorickshaw driver's association on the Mullaperiyar issue and the violence when a sub-inspector of police, who was passing that way on a motorcycle, stopped by the auto stand and told the auto stand leader to tell the cameras that Tamils would not rest until all Keralites are chased out of here.

'Nalla sollu ya. Be loud and clear, otherwise the Keralites will stop your water and trample you,' he said, before resuming his journey. The sub-inspector's comment cheered the auto stand members, who started hooting and hurling more abuses against Oomen Chandy and his team.

When I mentioned this to the SP, he said that he had not observed such a trend. 'It is possible that one or two elements could be there but they are present everywhere, and I did not face any such issues from my team,' he said.

After the situation returned to normalcy by end of January 2012, the top police officials in Chennai had asked the SP to prepare a report on what could have triggered the violence and if there were some organized groups behind it. 'We did try to probe on the root cause of the problem and if some extremist elements had infiltrated into the villages and triggered the protests, but so far, we have received no such evidence. It seems like a spontaneous outburst of violence,' he said.

Pravin Kumar pointed out that villagers stopped agitating only after they were assured that the water level would not be reduced below 136 feet in the dam and that they would get their usual quota of water. 'If the Kerala government tries to do something like that again, then violence will break out again, and it will be difficult to stop these people from crossing the border. These are poor villagers who have nothing to lose. They will go to any extent to fight for the water,' he said.

Before leaving, I asked him about the allegation that not many FIRs were registered in connection with the violence and that most victims of the looting and assault could not even register their complaints.

'We have registered at least a few dozen cases of vehicles being ransacked and homes being looted. But most victims had fled the district before these could happen and, hence, they have not filed a complaint. We are looking into all the complaints and probing. But it could take time,' he said.

I asked him about my contact Kumar whose Tata Sumo was one of the vehicles that were damaged which I had personally inspected. 'I just spoke to him and he said that Cumbum police did not take his complaint,' I said.

'I do not know about individual cases, boss. But you ask him to come to my office; I take grievance petitions from the public on a regular basis,' he said.

*

About 20 kilometres away from Madurai city along the Madurai–Bengaluru National Highway is the tiny village of Vadipatti. Surrounded by lush green paddy fields interspersed with occasional patches of vegetable gardens that included potatoes, onions and other crops, the entire village is barely a kilometre long with modest one-floor homes on either side of the only tar road that passes through the village.

I visited Vadipatti to interview Ramaswamy, who is an active member in the Periyar–Vaigai Farmers' Association in the Madurai belt. Ramaswamy is a perfect stereotype for a farmer—modest, docile and ageing. When he came to greet me, he wore a dirty white dhoti and a white shirt. We shook hands; he apologized for the grey stubble that had shrouded his face and invited me inside his house. 'This is the annual temple festival season and I am on a fast,' he said. Like most other families of farmers in the region, the youngsters in his home too had moved out to cities and bigger towns in search of employment, leaving behind the parents and grandparents to do the farm work.

As we chatted about the Mullaperiyar dam, the condition of farmers, and the future of the region, two of his neighbours who belong to traditional farming families joined us. Vadipatti is located approximately 100 kilometres away from the Mullaperiyar dam, and yet the profound impact that the Periyar waters have made to the region is obvious.

'To understand the impact that the Periyar waters have had on this region, all you have to do is compare the lands on our side of the highway with that on the other side of the Madurai–Bengaluru Highway. The Periyar Vaigai Irrigation canal passes only through this side of the road, irrigating all farmlands between here and the Vaigai basin on the other side. All excess water that we do not use will flow to the Vaigai basin and does not get wasted,' said Ramaswamy. 'While the ground water level is at 40 feet on our side, even if you dig a bore for 200 feet, it is hard to find water in the villages on the other side of the canal,' he said.

Ramaswamy and his generation of farmers took up the profession during the late 1970s and early 1980s and have never enjoyed the benefits of the dam even when the storage level was at 152 feet. 'Since we started working, the water level has been at 136 feet and there has always been a dispute of some kind between the two states,' he said.

Despite the reduced supply of water, Ramaswamy and other farmers in Vadipatti and surrounding areas have two crop seasons on most years, and rice and sugarcane are the chief crops raised here, too, like in most of the Periyar–Vaigai basin. While the Cumbum region gets water from June first week or so, water reaches the canals supplying to Vadipatti, Melur, etc. in Madurai district only after the water level in the Mullaperiyar dam reaches 120 feet. On some years, they receive water as early as mid-June, but there have been years when they got the first supply of Periyar waters only in late August and September. 'During such years, we restrict ourselves to one harvest season,' said Vedakan, another farmer from Vadipatti who took up agriculture after his retirement. He used to work as a schoolteacher during his younger days.

Approximately 1.4 lakh acres of land are irrigated in Madurai, Thirumangalam and Melur areas, out of which at least 55,000–60,000 acres have two crop seasons in a year, thanks to a steady supply of water for irrigation activities for ten months in a year.

Besides agricultural activities, the Periyar waters are also treated and used for drinking purposes in lakhs of homes in the region, as there is no other alternate source of drinking water.

While rice and sugarcane, both water intensive crops, are mainly raised in the Periyar irrigation area, the other regions that do not fall under the scheme continue to depend on the unpredictable rainfall, and restrict themselves to dry crops such as corn or even vegetable crops that can be grown in a short duration and in a relatively lesser time period with little or no rainfall. 'But now, those landowners are turning out to be lucky. Many of the lands on that side have been converted into plots and sold for residential purposes. We cannot even do that as these areas are classified as agricultural land and, due to the constant supply of water for so many years, the top soil is not suitable for construction activities. We cannot even change the crops that we grow as, if there is sufficient rain during that particular year, water would continue to flow into our farmlands automatically,' Ramaswamy said.

Unlike this generation of farmers who are well aware of the water woes of the region and the long, legal battle that has been going on between TN and Kerala over the Mullaperiyar waters, the previous generation and the generation before that have never had to worry about water supply and focussed only on their crops. Many of the present-day farming families have moved from other parts of the state to the Periyar–Vaigai basin after purchasing lands here only became they were assured of a steady supply of water and, thus, a guaranteed harvest.

If, for some reason, the Periyar waters stop flowing to their farmlands, the families here would be staring at a bleak future. 'We have been farmers for so many generations and do not know how people survived when there was no steady agriculture here. If this dispute is going to continue for more time, we do not know what else to do but do everything in our capacity to protect our waters,' Vedakan said.

Any move to reduce the supply of Periyar waters or totally stop it would not just be dislocating these people and rendering lakhs of farm labourers jobless, but a tradition and culture that has been built over a hundred years could get wiped away if the SC verdict goes against TN.

'Already so many changes have come across in this region due to the reduction in workforce available for agriculture. For an agrarian village in south India where cows were a part and parcel of our families, hardly any farmer owns cows or oxen in Vadipatti. The sad truth is that we even buy milk for our daily use from outside,' Vedakan said.

Besides, machinery has replaced a lot of farm-related activities here, too, which has put a number of small farmers in the dock. Expensive equipment for sowing and harvesting purposes cost lakhs of rupees and are affordable only for farmers who till large areas of land. Smaller farmers who own just an acre or two cannot afford this equipment and suffer most due to shortage of seasoned hands for labour. Over 70 per cent of agricultural land in the Periyar–Vaigai irrigated area belongs to small farmers, and their only solace these days is that they do not have to pump out groundwater and use electricity to irrigate their land. In the event of a reduction in supply of Periyar waters, they might have to depend on the depleting groundwater sources that could increase their expenditure significantly.

One of the first petitioners at the Madras High Court demanding the increase in the Mullaperiyar water level to 152 feet was a senior farmer Seeman from Melur. Janatha party leader Subramanya Swamy's petition for the same cause was after Seeman had filed, and since then, members of the farmers' association here have been seriously campaigning for their cause. Around thirty of their members even went to New Delhi and met with Dr Manmohan Singh, then prime minister, a few years ago to highlight their plight.

Just as is the case with the villagers who live downstream Periyar, the disgruntlement of the farmers with the state politicians is obvious. 'Successive political parties have surrendered our genuine right to more water that we won through years of legal battle to gain petty political points. A Supreme Court verdict is present to raise the Periyar water storage to 142 feet. Why has it not been raised yet?' asked Ramaswamy. 'While we are happy that the recent report of the Empowered Committee headed by Justice Anand has come in our favour, we are not hopeful about its outcome. Kerala will once again do something to prevent us from raising the dam storage level by hook or by crook.'

Towards the end of my interview, our conversation drifted to the recent riots that broke out in Cumbum and I asked them if they had participated.

'Yes, we went all the way from here to the Cumbum border and joined the protests. Every day, I went along with all the men from this village. We are watching every move of theirs and there will be another, much bigger riot when Kerala tries to misbehave with us. If the Mullaperiyar water does not flow into Tamil Nadu, not a single Malayali will stay in Tamil Nadu, we will drive all of them out,' Ramaswamy said.

Then he paused for a while and asked me what my native place was. I told him that I was from Chennai. He continued, 'We have been deprived of our quota of water up to 152 feet for the last thirty years despite a Supreme Court order. No Tamil politician is willing to take it up. Now, the empowered committee has again ruled in our favour and said that the water level can be raised to 142 feet. If Kerala doesn't respect the Supreme Court order this time, then we will also not respect the States Reorganization Act and capture the territory where the Mullaperiyar is situated. It should have come to Tamil Nadu as 60 per cent of the population in Idukki district is Tamil. If they don't want to respect the law, then we will also take it in our own hands,' he said, with an aggression that I never thought was inside him.

*

K.M. Abbas, former president of the Periyar–Vaigai Farmers' Association, lives with his extended family at their home in Cumbum. To most farmers in the Periyar–Vaigai basin, Abbas is like a father and the younger generation farmers affectionately call him Appa. For the last thirty years, Abbas has been heading the farmers' association in this region and has been spearheading

TN's campaign pressing for more water to the farmers and raising the storage level in the dam to 152 feet as per its design. To the proponents of a new dam in Kerala, Abbas is enemy number one.

When I had gone to meet him at his home in Cumbum, Abbas was out of station and staying with his close relatives in Kodaikanal, the hill station nearest to Madurai, to spend the summer vacation with his grandchildren.

'I am sitting here under a tree after my morning walk in Kodaikanal, but my thoughts are all about Cumbum and the progress of the case in the Supreme Court,' he told me over the phone.

While most farmers in the basin have taken a more radical stand about the TN–Kerala relationship over the Mullaperiyar issue, Abbas is a moderate face of the protestors. 'Tamils and Keralites have been living peacefully for generations before and after the dam came into existence. We have no animosity against Keralites and they are welcome to stay amid us. Our ire is only against the politicians in Kerala who continue to spread blatant lies about the safety of the dam and create a panicky situation. The Periyar–Vaigai basin farmers have no ill feelings against anyone.' He spoke in a tone that seemed like he was orating at a public function.

Abbas pointed out that while Kerala has been claiming for the last thirty-three years (since 1979) that the dam was unsafe and lakhs of people would die, not a single person had lost his life due to the waters released from the Mullaperiyar dam. 'What scientific basis does Kerala have to state that the dam is unsafe? All the technical experts have confirmed that the dam is safe and that there is no need to panic. Still, for thirty years we have been fighting to prove a fact that needs no proof and is obvious,' he said. 'Anyone who says that the dam is safe in Kerala is being attacked. Take the case of Professor C.P. Roy who recently came out stating that digging a tunnel at 70 feet would ensure safety of the dam. His home was attacked by goons. Even eminent Supreme Court judge, K.T. Thomas, who said the dam was safe has been threatened. If this is the case, how can people speak the truth?'

Until a few months ago, Mr Abbas was the president of the

association and an active campaigner when the riots took place. His sudden change of stance in stepping down as the president unilaterally raised a few eyebrows among farmers here who thought he could have been bought out. People close to him say that Abbas was advised by his family and friends to keep low after a group of strangers barged into his home and threatened to cause harm to his family if he continued to campaign against Keralites. 'But he is still the leader of this organization and nobody can forget the meticulous work he has done in collecting all details about the issue since 1979,' said a close associate.

Abbas and other members of the Periyar–Vaigai Farmers' Association have taken up their cause in several forums and have made representations at the highest level. A few years ago, around thirty members of the association went along with Abbas to New Delhi to meet the prime minister. They personally met Dr Manmohan Singh and apprised him of the present situation in the Periyar–Vaigai basin, seeking his help in resolving the dispute. 'But despite all these measures, we are back at square one because of the adamant stand taken by the neighbouring state,' Abbas said.

When I asked him on what would be a possible solution for the problem that was acceptable to the farmers of the Periyar–Vaigai basin who would be worst affected if there was a new dam, he laid down a list of conditions that need to be fulfilled before they could even think of a solution.

'As of now, we cannot even think of a solution until the Kerala government ensures that they are willing to arrive at a solution. To prove that, they have to first repeal the amendment to the Kerala Irrigation and Water Conservation Act 2006 that capped the maximum water level at Mullaperiyar to 136 feet. Let them allow Tamil Nadu to implement the 2006 Supreme Court verdict and raise the water level in Mullaperiyar dam to 142 feet before any future solution can be arrived at,' he said.

The Periyar–Vaigai Farmers' Association also wants the state government of Kerala to stop talking about a new dam when the case is still pending before the court. 'Even as a dispute is on, the state government has allotted over ₹500 crore to construct a new dam, violating all norms. First, they should stop indulging in such activities and be willing to cooperate with the TN government and the Central Water Commission in arriving at a solution. If the Kerala state government does all this, then we will be able to arrive at a peaceful solution to the problem, as per the directives of the Supreme Court,' he said.

*

On a normal day, retired PWD engineer R.V.S. Vijayakumar is a jovial person who has the amazing ability to see even the most humiliating situations in a lighter vein. This personality trait has greatly helped him in dealing with several tense situations while he was the chief engineer of the Madurai Circle of TN PWD and was being harassed by the Kerala police who are providing security for the dam. The man, who is in his sixties, is an expert structural engineer with a masters in engineering from a reputed university in the state, and has been at the helm of affairs in the Periyar–Vaigai project for several years.

Since his retirement a few years ago, Vijayakumar has been an active member of the senior PWD engineers association and is currently its president. Apart from the Periyar–Vaigai Farmers' Association, the only other organization that has been actively campaigning for the cause of the TN government and its farmers is the senior PWD engineers' association.

I met Vijayakumar at his home office in suburban Madurai on a Sunday afternoon. The former chief engineer, who was neatly dressed in formal clothes, was seated behind a large table with piles of files and scribbling pads strewn around. Above his chair was a large photograph of V. Prabhakaran, the deceased leader of the Liberation Tigers of Tamil Eelam, holding a machine gun, wearing the military uniform that he was well known for.

'The main problem with the Mullaiperiyar issue is that no politician in Tamil Nadu has the guts to ask Kerala to play by the rules. They are not willing to listen to the Central Water Commission or the Supreme Court and are acting on their own. But our politicians are too busy to look into these issues and keep harping, time and again, that we have to try to resolve this issue through dialogue,' he said, giggling intermittently all through the conversation.

'You might find it funny but I can assure you that most politicians who become PWD ministers in Tamil Nadu do not even know whether the Mullaiperiyar dam is in Tamil Nadu or Kerala,' he continued. 'The Supreme Court verdict in 2006 that came in our favour was during my tenure as the chief engineer of Madurai. When I got the news from Delhi, I went to a senior cabinet minister and told him that we will be able to give the Tamil Nadu farmers a pongal gift as the verdict seems to be in their favour. The minister laughed at me and said that I was dreaming. He told me it was impossible for us to get more water.'

Mr Vijayakumar has been a part of several meetings between the two state chief ministers organized by the CWC and considers such efforts as a mere joke. 'We would all go, right from chief minister, PWD minister, PWD secretary and a bunch of engineers from both states. Talk some useless matters, have lunch and again give boring lectures on interstate cooperation and return home,' he said.

According to him, TN can get its due of Mullaperiyar waters after raising the storage level in the dam to 152 feet only through tough negotiations. 'They do not respect any other stand. Let us block all highways to Kerala. Almost all food products and natural produce reach Kerala only from Tamil Nadu. Let us resort to an economic blockade and then discuss on the safety of the dam. The state that does not respect the Supreme Court verdict or the opinion of some of the smartest engineers in the country needs some arm twisting to be done to have our interests protected,' he said. The jovial mood gave way to anger and frustration.

Over the last several years, the Senior TN PWD Engineers' Association has been actively campaigning for raising the storage level in the dam as well as for protecting the interests of farmers groups in the Madurai region. Since a large part of the campaign for a new dam has been done through the media in the form of newspaper articles, short films, etc., the senior engineers association has taken up the task of providing information to media houses in TN and Kerala to counter the claims made by the Kerala government.

The association has even made a short film on the strengthening measures that have been taken up by the TN PWD officials following the CWC recommendations and discussing the safety factors of the dam. 'We had put in ₹1 lakh from our pockets to make that short film. It is available on the internet for anyone who intends to know the reality behind the strength of the dam,' Vijayakumar said. The short film was mainly made to educate the member of the EC chosen by the TN government a few years ago. When the senior engineers found that the EC member, like many important people in the state, was unaware of the facts behind the Mullaperiyar dam and the hard work done by TN PWD engineers in maintaining the health of the construction in the best condition, they spent their money to make the short film and release it. Several booklets have also been published in Tamil, English as well as Malayalam to be distributed to the public in both TN and Kerala by the association over the past few years.

Mr Vijayakumar and other senior PWD engineers also organized several public meetings in Madurai and the rest of TN to highlight the misinterpretation of facts (as they call it) by the Kerala government and media in the Mullaperiyar dam row, and have quite a following in the region. He feels that the SC or the central government cannot bring about a solution acceptable to both sides, and that the public is losing faith in the state and central governments.

'The people in Delhi do not know the situation here. A portion of Tamil Nadu is turning into another Kashmir. The Tamils are agitated and seek justice. We have been denied our genuine right to water everywhere. Be it the Cauvery River or the Periyar or the Krishna River, Tamils are always at the receiving end. If the Periyar water does not flow into Tamil Nadu, this land will embrace a civil strife. Let us choke the Keralites by shutting down all the roads to the state. All essential supplies are reaching Kerala only through TN and from this region. We have to cut all economic trade, chase out the thousands of students from Kerala, and ensure that no Malayali establishment is here, if they show their pranks in the Mullaperiyar dam. They don't know but this land will become a breeding ground for terrorists,' he said. 'I have been often asked by agitated groups of youth on what to do. They ask me if they need to ransack homes, march to the dam and take control, or start attacking Keralites. People here are ready to go to any extent to protect the Mullaperiyar dam,' he said.

*

7
Politics and Polemics

Despite the terrible social, environmental and economic record of large dams and the many other ways for satisfying the need for energy and for managing land and water, huge dam projects continue to be proposed and built. The dam industry juggernaut maintains its momentum because constructing dams benefits powerful political and economic interests, and because the process of planning, promoting and building dams is usually secretive, and insulated from democratic dissent. Those who suffer because of dams—whether directly through the loss of their livelihoods or indirectly through government subsidies to uneconomic projects— are rarely able to hold dam building bureaucrats and consultants accountable for their actions. The lack of accountability is clearly at its worst under authoritarian regimes and where democracy and the structures of civil society are weak. But even in supposedly advanced democracies, dam building agencies have for years insulated themselves from public control and avoided independent scrutiny of their assumptions to justify projects. —***Patrick McCully***

Silenced Rivers: the Ecology and Politics of Large Dams

If there is any aspect of the Mullaperiyar controversy that the people of TN and Kerala agree upon, it is the staunch belief that their politicians have deceived them time and again.

Irrespective of whether the ruling party is the All India Anna

Dravida Munnetra Kazhagam (AIADMK) or the Dravida Munnetra Kazhagam (DMK) in TN or the UDF and the LDF in Kerala, the only reaction that one evokes from common people affected by the dam row, while talking about their politicians, is sheer contempt.

And they have their reasons. Despite the dispute over safety of the dam and fixation of water level being at least three decades old, neither the two top Dravidian parties in TN nor the two major fronts in Kerala, who have been shuttling in and out of power with amazing consistency, have done anything to solve the issue. The dam has been around for 119 years and there have been complaints about the safety of the dam since 1906. Almost every year, with the exception of drought periods, the Mullaperiyar catchment receives good rainfall and the dam is filled up to its brim, that is, 136 feet fixed as the FRL. How is it that the issue hogs the limelight only during a few years, and that too for just a brief period, and then vanishes as mysteriously as it erupted?

To understand the complex politics weighing over the issue and the people's interpretation of the political moves in either state, let us examine the most recent escalation of violence during late 2011.

According to newspaper reports in Kerala, the most imminent cause for the protests gathering momentum during November 2011 were the frequent tremors that were recorded in the region at a time when the Mullaperiyar dam was filled with water. The Mullaperiyar Samara Samiti had been holding chain protests for six years and there had been various other groups that had been agitating even before that. The outburst of violence, according to the protestors, was due to the earthquakes that triggered panic among the public.

But, for the people of TN, the entire controversy that escalated into a full-fledged riot and economic blockade along the interstate border was caused by the screening of a film called Dam 999, made by an amateur film maker Sohan Ray. Despite the movie not having anything to do with the Mullaperiyar dam or any existing dam but a mere fictional exploration on a dam disaster, the TN state government deemed it fit to ban the film in the state and aroused the curiosity of the Tamils.

Until the film was banned, people in most parts of TN had no idea of a brewing contempt in Kerala against the TN stand on Mullaperiyar. Similarly, life was as usual for the villagers and plantation workers in Theni district until the first signals of dissent came in the form of hate messages on TN-registration government buses about ten days before the riots broke out. Could there be other reasons for the timing of these riots?

When the Mullaperiyar controversy was at its peak, several political pundits and analysts came up with various theories on the timing of the conflict that arose out of nowhere and threatened interstate relations like never before.

'One possible connection for the timing of the riots was the by election in Pirovam constituency that was scheduled to be held in March 2012. Faced with an electoral battle and no powerful ammunition, the ruling UDF could have stirred the safety of the public in Kerala and their stand on insisting for a new dam, thereby protecting the lives of Keralites. The politicians could have used this as a platform to express their solidarity with the people in building a new dam, after decommissioning the existing one,' says a senior Trivandrum-based journalist.

For the record, Piravom town is located way below in Muvattupuzha taluk, Ernakulam district, and is barely 33 kilometres away from Kochi. To imagine that the Mullaperiyar waters would wash away this town is nothing but exaggerated imagination. 'However, it has been a common site in Kerala for all political parties and outfits to raise their pitch on the Mullaperiyar issue during elections, as has been done for the last several decades,' said a senior journalist with a Kerala-based English daily.

The timing of whipping up passions in neighbouring TN during the same period is even more amazing. Until November 2011, the entire state's attention was diverted further down south near the Kanyakumari coast, where a tiny group of fishermen from the Idinthakarai village had successfully stalled the launch of a ₹14,000 crore nuclear power plant in Koodankulam through peaceful protests with support from anti-nuclear activists.

These protests that had been gathering momentum on a steady basis since early August had just begun to embarrass the state government which was already reeling under a severe power crisis. Over the next few weeks, the anti-nuclear protest had gathered so much momentum that work at the Koodankulam was stopped for almost three months as the protestors blocked roads and prevented workers from entering the premises. Many believe the Mullaperiyar issue was used by the TN government to divert media glare from the embarrassingly powerful protests.

The common public in Chennai and Trivandrum as well as the rest of the two states see the rising tempers over the Mullaperiyar dam as a phantom created by the state government to put the Koodankulam protests on the sidelines and shift public focus elsewhere.

Political analysts have also pointed out that this diversionary tactic has worked excellently in the most recent case. During the Piravom bye election that took place in March 2012, UDF (the ruling front in Kerala) marched to a thumping victory, while around the same time, the TN government headed by the AIADMK managed to silence the anti-nuclear protestors at Koodankulam and resumed operation of the nuclear power plant.

People have now forgotten the controversy and the violence and moved on with their lives. But the Mullaperiyar dam continues to remain as it was, vulnerable or not, as during November 2011. The region is still in seismic zone 3 and the water is filling up fast in the catchment again. Yet, the dam row phantom has gone back to its dormant state only to raise its ugly head whenever its political masters want.

*

The incidents that took place in 2006 following the SC verdict that came out in favour of TN allowing the FRL in the dam to be raised to 142 feet, and the moves made by the top politicos in both states, are a classic example of how politicized the dam row is.

The SC came out with its judgment on February 2006 allowing TN to raise the water level. The AIADMK government was in power at the time and the elections were just round the corner and due in May that year.

'For a water-starved state where the farmers in the Madurai–Theni belt have been yearning for more water and have been fiercely fighting for the Periyar waters, the state government should have jumped at the opportunity and asked TN PWD officials to put the spillway shutters and raise the water level. But that did not happen,' said a former PWD chief engineer, who was at the helm of affairs during the period.

'I went and told the second-in-command at the state cabinet of the announcement of the SC verdict in our favour and that we needed to implement it immediately. But the minister scoffed at me and said that the party had decided not to make any decision on raising the shutters until the Assembly elections were over. I was so agitated as we had suffered so much at the hands of the Kerala police and forest officials for years. Now, even when the SC has ruled in our favour, politicians are not interested in following the guidelines for their petty gains,' he said.

The potency of the verdict remained for barely twenty days or so. By mid March, all the politicians in Kerala set aside their differences and in a rare show of unity, passed the amendment to the Kerala Irrigation Act, as mentioned in detail earlier. The audacity of a state cabinet to pass an Act in direct violation of the SC verdict has not just made matters worse for the state government but has also bred hatred among Tamils for their neighbours, as they found the Kerala politicians to be extremely cunning while the Tamil politicians turned out to be an impotent lot incapable of even implementing the SC order.

Meanwhile, the ruling party in TN was busy campaigning for the elections, and lost the government. The DMK-led coalition that came to power in 2006 and ruled the state for the next five years could not implement the order, and instead took the matter again to the SC. These activities showed both state governments in bad light and the public lost trust in them.

This inability of the TN politicians in implementing the SC verdict that was in their favour has been seen as a major political victory by the LDF in Kerala, that was ruling at the time. A former LDF minister even said that the DMK lost southern TN in 2011 assembly elections mainly due to its inability to raise the water level to 142 feet, despite having the SC order in their favour.

Even before and after these incidents and sporadic outbursts of violence, several meetings have taken place at the official level between the chief ministers of TN and Kerala mediated by the CWC, but have yielded no results. Following the complications that arose following the SC verdict, the courts advised both the states to discuss and arrive at a political solution yet again. Successive meetings between chief ministers have also taken place and they have shaken hands, but the issue has not even moved one step closer to resolution.

*

Of the several dozen politicians in Kerala who have visited Chappathu, Vallakadavu and other villages situated along the Periyar urging the people to fight for a new dam, only a few have managed to win over the trust of the people. Comrade N.K. Premachandran is one of these.

The former minister for water resources, who was instrumental in ensuring that the SC verdict was not implemented by TN in 2006 and that the status quo was continued over the water level in the dam, is probably the only politician who remains in the good books of the public in Idukki district, and has been seen as being genuinely involved in the fight for a new dam for the last several years.

When I met Premachandran during May 2012, he was busy preparing for hectic campaigning for a coming bye election and was at his home in Kollam. Despite his hectic schedule, he agreed to meet me at his sprawling bungalow, barely a few kilometres away from the Kollam Railway station, for a freewheeling interview.

'The basic issue concerning the Mullaperiyar dam is that people living downstream of the dam are apprehensive that the dam is unsafe and could break any time. Despite the strengthening measures and the series of tests conducted

by the central agencies, the authorities have not been able to
wipe away that fear from the minds of people. As politicians,
it is our job to protect and secure the lives of our people and
represent their fears,' Premachandran said. 'When there is
a genuine apprehension, it is our duty to address it.'

Elaborating the stand taken by the Kerala government,
Premachandran said that since the dam was then 117 years old
and people were living in fear, it was probably a good time to build
a new dam in its place. 'It is not going to last forever anyway. We
might as well build it now, so that these people living under the dam
can feel safe. We have never said that Tamil Nadu will not get any
water. The neighbouring state has been getting the Mullaperiyar
dam waters for over a hundred years and will continue to receive
it. But the safety of our people is our primary concern,' he said.

However, Premachandran was apprehensive about the
feasibility of the task. 'It is all politics. At least in Kerala, the UDF
and LDF more or less have the same stand in Mullaperiyar and
have a common agenda. But look at Tamil Nadu, their politics is
so closely fought. Any party that supports the construction of a
new dam is doomed and has no political future. Then, there are
also people like Vaiko who have taken an extremist stand. In this
situation, I don't see a political consensus over the issue, at least in
Tamil Nadu,' he said.

While Premachandran has been hailed as one of the few
politicians in Kerala who is concerned about the people, on the
other side of the border he is seen as the chief perpetrator of lies
and is held responsible for the 2006 fiasco. 'Come every winter, our
Thozhar (comrade) Premachandran will come to Idukki with his
bagful of lies and start false propaganda about the lack of safety
of the dam and the selfish Tamils who take away their water. It is
only people like this man who mislead the Keralites, and are even
willing to go to the extent of spoiling the decades-old bonhomie
between the two states,' said a retired PWD engineer and an active
campaigner for the dam safety during a public meeting held in
Madurai district to much applause. Bashing Premachandran and
former chief minister Achuthananthan is a favourite indulgence

for many in TN when it concerns the Mullaperiyar dam.

But the young Marxist leader does not understand the hatred that the Tamilians have inculcated for him. 'We are living in a democratic era where people have a right to express their woes. Kerala, too, has water issues and is starved of drinking water for six months in a year. Yes, we will give water to Tamil Nadu but we need some water too, and we cannot keep resting on an agreement written at a time when the politics and governance in the country was entirely different. A new dam is needed and along with it, a new agreement that adheres to modern principles of justice that is negotiated between the two states, using the centre as a mediator, is also needed. It is only fair that we ask for a new dam and new contract; we never denied water to Tamil Nadu,' he said.

Unlike most other political leaders who do not know the complexities involved in the dam row and haven't the slightest clue about the CWC reports and its intricacies, this leader has made an effort to study the Mullaperiyar issue in detail, and claims that the report filed by the Justice Anand committee as well as by the earlier committees had several flaws in them. He blames the UDF government for not having raised the flaws in the earlier safety report that eventually led to the case tilting in favour of TN.

'The safety tests conducted by the members of the Empowered Committee have several issues. For instance, the calculation of the Probable Maximum Flood level has not been in accordance with factual data available with the Tamil Nadu PWD. The technical committee members, Thatte and Mehta, have been involved in the previous committee testing too. How will they go against their own conclusions? Hence, to accommodate their earlier findings, several misinterpretations have been made in the report, and we will fight it in court,' Premachandran said, as he recounted several more facts and figures to emphasize his point.

The former minister does believe that neither the central government nor the CWC has been fair to Kerala. The repeated assurances by top dam experts from the union ministry of water resources on the safety of Mullaperiyar dam, according to him, are just being made to please TN, which has a bigger clout in central politics and has the financial power to influence the CWC.

It is a common feeling shared by the political fraternity in Kerala. They believe that TN, which is a much wealthier state with thirty-nine MPs, has traditionally continued to be able to swerve the central government decisions in its favour using political clout. 'Kerala, being a much smaller state with lesser money power, has always been at the receiving end of the central government policies. The centre has time and again failed to protect the interests of minority states,' said another prominent politician, while commenting on the politics behind the dam row.

He said that the ruling party in TN will do all it can to ensure that the Mullaperiyar dam does not get decommissioned during its tenure, as it would mean doom for the political party in the region. 'To avoid the decommissioning, TN politicians would go to any extent, and they have already been pumping in a lot of money to quite a number of people. How else do we explain some of the sudden changes in the stand of some key members of the protest groups? Even the empowered committee member appointed by the state of Kerala claims that he does not operate in the interest of his state in an open forum,' the Congress party politician said.

So, do they even foresee a political solution for the problem?

'Yes,' says Premachandran. 'It is possible, but only a very powerful and rational leader in Tamil Nadu can ensure that the Mullaperiyar dam row ends. We need a national-level leader from TN, who has enough popularity among his people, to take a stand considering national interests even at the risk of receiving brickbats from his political rivals within the state. Such a leader is not available in the current political scenario but we cannot rule out that possibility in the future,' he says.

But the politicians in Kerala, be it the LDF or UDF, are united in their stand that the safety of their people is of utmost importance. 'If they will feel safe only when there is a new dam, then we will continue to fight for it,' said Premachandran.

*

Politics in the Dravidian heartland of TN has never been conducted the way it has been in the rest of the country. For the past four decades, no national political party has made any significant inroads in the dry and dusty southern part of the country since the Dravidian movement gained significance. Both the AIADMK and the DMK always bet on development platforms, and their contribution is evident in the progress made by the state when compared with the rest of India. TN is widely considered today as a manufacturing hub, while IT and many other sunrise industries have also found a home. But the state has always been deprived of water.

Geographically located on the leeward side of the Western Ghats, TN does not have many rivers passing through it. Even the few rivers that flow east are not perennial, and several interior areas are still water-starved despite the progress. The state already had ongoing water disputes with all the other southern, states including over the Cauvery in Karnataka and the Krishna in Andhra Pradesh. Water issues dominate local politics in the state, and any party that does not bring water into the state for the starving farmers does not have any chance to succeed in capturing power in TN.

The Mullaperiyar waters feed around five districts in the state, which includes five parliamentary constituencies and more than twenty-five assembly constituencies. Any politician who ignores the livelihood concerns of the Periyar basin farmers is not likely to receive even a single vote from each of the 500-odd villages that are present in the region.

Although the present chief minister Jayalalithaa and the former chief minister Kalaignar have dominated the vote bank in this part of TN, the only politician whom the farmers hold in high regard when it comes to water dispute with Kerala is Vaiko, a.k.a Vai Gopalasamy, leader of the MDMK.

When the December 2011 riots was at its peak, the only politician who was allowed inside the region by the protestors was Vaiko. Even senior AIADMK leaders like O. Paneerselvam, who belongs to the region, was shown a slipper when he tried to address a public meeting in Theni district during the period.

Vaiko firmly believes that the Mullaperiyar dam is one of the strongest in the country and it has been the heart of his campaign. 'I can personally vouch for it,' he says.

For the last several years, Vaiko has been actively campaigning for TN's cause in the Mullaperiyar dam row, and has even gone to the extent of publicly endorsing an extremist stand over the dam row. In several public meetings held in Madurai and other areas, Vaiko and his supporters have demanded for the state government to block all roads leading to the neighbouring state and stop exports of vegetables, meat and several other essentials from TN.

Vaiko explains that his hard stand is to counter the false propaganda that successive Kerala state governments have indulged in over the past several years. 'Even as early as 1979, this propaganda machinery was active and an article in *Malayala Manorama* raised questions about the Mullaperiyar dam at the time,' he said. 'A panel of experts from the Central Water Commission said that the dam was safe. Still, they advocated strengthening measures to ensure more safety and requested us to reduce the water level to 136 feet until the measures were completed. But our PWD engineers were never allowed to complete the work peacefully as Kerala politicians and bureaucrats tried to stop the work in some way or another. Although we have completed all the works after two decades or so, the dam level still remains at 136 feet and we have been able to do nothing about it. Every year, that amounts to a loss of about ₹130 crore to the TN government and its people in terms of reduced power generation as well as availability of water for farming. I am just taking these facts to the people as they have a right to know.'

When TN PWD officials were not allowed to raise the water level back to 152 feet, the farmers went to the court, and the legal battle that ensued turned out to be in favour of TN. 'Despite two independent committees of experts certifying that the dam is safe, their politicians say Keralites are still apprehensive and project a doomsday scenario of a large part of the state getting washed away if the Mullaperiyar breaks. It is all humbug. Even if the dam does break, the water will be stored by the Idukki dam, as several people have already stated. Moreover, those living along the river need not fear as they are staying at a higher elevation,' he says.

One of the biggest failures of TN politicians in the Mullaperiyar row has been their inability to implement the verdict of the SC and raise the water level to 142 feet in 2006. At the time when the verdict was announced, the farmers in Madurai and Theni celebrated it as a great victory, but the ruling government at the time decided to play it easy as the state assembly elections were just around the corner.

When asked to comment on that fiasco, Vaiko says that politicians in TN are seldom concerned about their people. 'They are only bothered about protecting their interests. We had betrayed our people despite being legal victors. But that's not the whole story. The Supreme Court verdict was overruled by a bill passed in the Kerala State Assembly within twenty days of the verdict. It is a constitutional failure and challenges the sovereignty of the country as well as the supremacy of the Supreme Court. But the central government and the courts decided not to react to such an outrageous move. Instead of nullifying the act, they just let them get away with it. When I question this, people call me an extremist,' he says.

Until then, the Mullaperiyar dam was not present in the list of dams in Kerala. Immediately after the SC verdict, the Kerala government had listed the Mullaperiyar as one of the dams in Kerala and fixed its maximum water level at 136 feet. Vaiko says that the Kerala government had been demanding for the decommissioning of the dam and the construction of a new one only after 2006. 'All my campaign has been to counter that move. We have travelled to over 538 villages in the Periyar–Vaigai basin and met with the farmers personally to apprise them of the situation. The uprising that we saw last year did not happen on its own. It was the result of years of work done by us and the farmers' association in enlightening the situation to the poor folk,' he says.

Moving on to the environmental violations, Vaiko pointed out that when the Kerala government had pressed for a new dam and sought environmental clearance to conduct a feasibility study for the new dam site, the union government had once again betrayed

TN. 'We are living in an age when the Supreme Court is even planning to ban tourism in tiger reserves to protect the animal and its habitat. Despite the presence of several prominent ministers from TN including Raja and T.R. Baalu in the cabinet, we could not prevent the then environment minister Jairam Ramesh from giving a clearance to conduct a feasibility study, which is a gross violation of all environmental norms. How can one take a soft approach when the central government is so biased against our interests?' asks Vaiko.

Between 2006 and 2011, the Kerala chief minister has met the prime minister to press for the state's case on three occasions, but the then TN chief minister has done nothing, according to the MDMK leader.

'I would like to go on record and say the then Tamil Nadu chief minister Karunanidhi had betrayed the state in the dam row just to ensure that his cabinet positions in the centre were secure,' says Vaiko. To emphasize his point, he said that since the Left parties were in power in Kerala and were also a part of the union government coalition, Karunanidhi's DMK, being an active partner in the UPA-1 government, decided to play along with the CPM coalition government in Kerala, disregarding the interests of the Tamil people. 'This was the biggest flaw that has led to the present condition of animosity between the two states,' he says.

Vaiko's current preoccupation is the Dam Safety Bill that has now been mooted in Parliament. 'If the bill comes into force, each state will have the right to decide on the dams in their territory. For a state like Tamil Nadu, that is at the receiving end of all major rivers, this bill will be a disaster. The TN government should oppose it tooth and nail, and if they fail to do so, there will be a civil war in this country and we will be another Russia,' he says.

As for the Mullaperiyar dam, Vaiko firmly believes that if a new dam is constructed, then TN will not get a single drop of water. 'The site of the new dam is at a much lower elevation and we will not get any water in the foreseeable future. The only commitment that we, as politicians, owe to the Tamil people is that such a decision does not come into force during our lifetime, for it will be a betrayal to one's motherland and history will never forgive us,' he says.

*

A few days ago, when I asked TN-based senior political analyst Cho Ramasamy if a political solution was possible in the Mullaperiyar dam row, pat came the reply. 'What do you mean by a political solution? I do not understand, are you trying to say that the states should arrive at some kind of a compromise solution?' he asked.

Cho Ramasamy is in his late seventies. Over these last seven decades, he has donned several hats, including that of a movie star, journalist, political kingmaker, and has earned the reputation of being an expert at analysing the political scenario down South. For the last several years, he has been running a vernacular magazine *Tughlak* operating out of his modest office in Greenways Road, Chennai. An open supporter of Narendra Modi, Cho is one of those few who speaks his mind, say close friends and colleagues. I had interviewed Cho to understand the politics behind the dam row that had prevented a possible solution in the past as well as the near future.

When I told him that I do not believe that the present case proceeding in the SC, which will soon reach trial stage, might not give a solution after all, considering the fact the earlier judgment had just a lifespan of twenty-odd days, he replied that it was not the first such instance.

'Let's take this particular case; the Supreme Court had clearly mentioned that the dam is safe and water could be raised to 142 feet, but the Kerala government has blatantly violated the court. It is a failure of the constitutional machinery and ideally, the state government should have been immediately dismissed for not accepting the court verdict. But then, this is politics,' he said.

Cho went on to say that in the present scenario with the Congress-led UPA at the centre, ready to face the next big battle in 2014, and Congress being the ruling party in Kerala, no prime minister or central government would want to antagonize the state. 'It would mean that Congress party would lose all its votes in

Kerala. On the contrary, what does Congress have to lose in Tamil Nadu? After all, they have never managed to gain power over the last four decades or so and have only played second fiddle to DMK and AIADMK. As of now, Congress has much to lose in opposing Kerala than in opposing Tamil Nadu, hence, they would not act against Kerala government in any case,' he said.

The scenario was no different when Karnataka state government flouted the SC verdict in the Cauvery water dispute. 'It is not a new trend at all. Several state governments have flouted Supreme Court verdicts and acted on their own when it comes to water disputes. Unfortunately, the Supreme Court can only put orders but they have no machinery to ensure that it is implemented. That job is left for the central government and the centre does not operate by court verdicts but by the political compulsion of the time. So, it all eventually boils down to politics,' he said.

Unlike Vaiko, Cho refused to accept that the state governments had made a mistake in handling the Mullaperiyar issue. He is of the opinion that both the DMK and AIADMK had been doing their best to ensure that farmers' interests are protected through legal means. 'Even in 2006, neither DMK nor AIADMK could have done anything unless they got the signal from the centre. While the dam is in our control, security is provided by Kerala state. If the TN PWD engineers had tried to increase the water level, the police who are suppose to guard these engineers could have attacked them,' he said. 'Only a strong central government with an able prime minister will be able to rein in the states and ensure that the rule of law is followed. Until then, such issues will continue to crop up regularly. After all, when compared with Western democracies, we are still young and might take time to get mature enough to take political decisions on serious issues,' he said.

In a system where courts have to depend on an elected government to implement its verdicts, the judiciary in the country too sways to the mood of the public. 'The most recent report of the EC appointed by the Supreme Court is a classic example,' points out Cho Ramasamy. 'While all the experts have agreed that the dam is safe for the near future and there is no need for any panic, they still add a clause mentioning that a new dam could also be considered as an alternative in the long term. If the present dam is safe, then what is the need for such a statement? It is a clear indication that the courts are playing to the gallery and ensuring that nobody feels defeated. This is not the first instance when the courts play to the gallery,' he said.

In many ways, the Mullaperiyar dam controversy is symbolic of much of the malaise that affects the country as a whole. Opinions of technical experts have been ignored, verdicts delivered by the highest judicial body have been flouted at will, and politics has taken centre stage in an issue that directly affects the lives and livelihoods of lakhs of people.

*

8
Role of the Media in the Mullaperiyar Conflict

SOMETIME DURING MAY 2012, while engaged in the legwork for the book, I took a brief two-day vacation at Munnar, the scenic hill station in Idukki district, Kerala, to celebrate my wedding anniversary. On the last day of my trip, I finally caught up with a newspaper that was present in the reception area of the resort. It was a leading English daily printed in Kerala as well as TN. To my surprise, the Mullaperiyar story was on Page 1 as the EC appointed by the SC had tabled a 200-odd page report to the court and the conclusions of the report reached the media.

The front page story was published based on a statement issued by the Kerala government which claimed that the SC EC had given the green signal for a new dam 'with certain conditions'. I was shocked and wondered if I had to rush to Theni as I was certain that another riot would break out anytime now.

Later that night, I reached my home in Coimbatore and couldn't wait to grab the same newspaper (Coimbatore Edition) to read the full story. To my surprise, the Mullaperiyar-related story was on the front page of the Coimbatore edition too, but it stated that the SC had confirmed that the dam was safe and there was no need for a new dam. The last line of the copy mentioned that the EC had also suggested construction of a new dam as an alternative. The same newspaper had published two different reports based on the findings of the single committee giving two conflicting conclusions in two different editions on the same day. Ironically, only those few who travel across both states on the same day can enjoy such jokes.

Veteran journalists, however, find nothing wrong with the conflicting reports published by the same organization. 'The newspaper is just like any other product. It has a target audience and publishes news that is relevant to its readers and information that the audience is interested in,' says senior journalist R. Bhagwan Singh, consultant editor at *Deccan Chronicle*, Chennai. 'For instance, the report submitted by the EC has over 200 pages of content and states both. The committee has stated that the dam is safe and at the same time, it has said that a new dam could be an alternative long-term solution. In this case, both the editors who chose the news for the front page are right, as the report contains both the conclusions. The Kerala editor chooses what is relevant to his state while the Tamil Nadu-based editors choose what is relevant to TN,' he states.

A large part of the media hype over the Mullaperiyar issue has been due to the fact that while the dam is located in one state, its beneficiaries are in the rival state. While some of the apprehensions about the Mullaperiyar dam's safety might be true and have been adequately reported in Kerala, the tests done on the dam and the stringent maintenance standards followed by the TN PWD engineers are not accessible to Kerala journalists, as for all practical purposes, the dam does not belong to them. Hence, while the public in TN has been repeatedly fed on news that the dam is in good shape for whatever reasons, their counterparts in Kerala are fed exactly the opposite.

Over the years, the issue has become so entangled with local politics that it has become next to impossible, especially for local vernacular media, to publish a report or take an editorial stand that goes against their people.

Despite the presence of many national-level news magazines and television channels, a vast majority of the public in TN and Kerala still get their news from local newspapers and vernacular TV channels that are entirely operated within their state. Hence, public opinion is largely formed by what appears in these channels and print media to a great extent.

A senior journalist based in Trivandrum said that the Mullaperiyar issue has been hijacked by politicians and various groups for such a long time that any news item that is even slightly tilting towards the stand taken by TN will be deemed paid news. 'We certainly run that risk. The public has been so brainwashed that they do not even trust senior SC judges when they claim that the dam is safe. Any newspaper organization, especially vernacular news media, will run the risk of being branded as "sold out" to the neighbouring state if they even slightly slander the local public opinion on such sensitive issues,' he said.

*

If one were to go back into the role played by media over the last century, one has to go as far back as to the time when the dam was constructed. Records indicate that some of the first reports that were published questioning the safety of the dam came out in Kerala as early as 1925. During that period, reports that appeared in a section of the press in present day Kerala led the Superintending Engineer, Trichy, C.T. Mullings to rush to the dam site and check the stability of the structure. Following his investigation, the Government of Madras had issued a press release on December 22 that the reports about the leakage were misleading and that the British government was taking necessary steps to improve and stabilize the dam.

History tells us that every instance of stand-off between the two states over the dam was triggered by media reports. The floods of 1961 that inundated parts of Alwaye and Perumbavur triggered a report raising concerns of the dam safety during the following year, which led to inspections and tests, the results of which confirmed that the dam was safe.

The next time the dam safety became a concern was following the Morvi dam disaster in Gujarat during 1978–79. Again, the media in Kerala raised concerns about the safety of the dam and related the disaster in Gujarat with an imminent dam break in

Kerala, triggering mass hysteria among the public in Kerala.

Following these reports, the CWC intervened in the row and conducted a series of tests which eventually ended up strengthening the dam to the present safety standards. Even as the strengthening measures were being conducted over a period of twenty years, the issue regularly hogged the limelight and ensured that the people were always aware of what was going on.

After the case went to the court, every minor development was keenly monitored and adequately reported. Even adjournments and observations made by judges have been widely reported and reputed persons like senior engineers from the union ministry of water resources such as Dr Thatte and Dr Mehta, who were a part of the most recent EC, have been seen as evil for their role in emphasizing that the Mullaperiyar dam is safe.

'Politicians and bureaucrats have conveniently exploited the media to achieve their ends, and have consistently fed reporters with news that they intend to communicate with the public. In a democracy, it is a healthy trend that the media is watching controversial issues like a watchdog. Unfortunately, the flip side of it is that people are fed with much more, usually irrelevant, information than they need,' says the editor of a vernacular magazine in TN. 'It is up to the reader to discern the flooding of information and arrive at the truth. We are bound to convey what their leaders speak irrespective of their intent.'

Many in the media business even feel that the most recent violence that broke out at the TN–Kerala border was largely a creation of the media who are hungry for sensationalism. 'How else would one describe the sudden stirring up of emotions? Even now, a couple of newspaper reports are sufficient to trigger a fresh wave of violence between the two states and it will continue to remain that way,' says the magazine editor, seeking anonymity.

*

One of the few English newspapers that were truly at the receiving end of the protestors' ire in TN was *The Times of India*, Chennai edition. While many English language publications were slammed by several fringe political outfits and Tamil nationalist groups for making attempts to report the other side of the story, mischief mongers in these groups went to the extent of accusing *The Times of India* of being anti-Tamil.

During December 2011 and January 2012 when the riots were at their peak, several outfits had burnt copies of *The Times of India* at protest venues and accused its editorial, a large number of whom were natives of Kerala, of being disloyal to their profession and showing bias towards Kerala. Simple editorial decisions such as spelling the dam as Mullaperiyar (as is mentioned in the ministry of water resources website) as opposed to Mullaiperiyar, as is pronounced in Tamil, were being read as signs of loyalty to the rival state.

Jaya Menon, political editor of *The Times of India*, Chennai edition, rubbishes claims that the editorial policy was tilted towards Kerala. 'It is just propaganda made by some political outfits to gain mileage. We were just as free and fair in reporting events as they had unveiled. Our reporters on the field were in regular touch with us and we decided on printing news as it happened,' she says. 'Yes, there were some people who were spreading rumours about our news articles not being fair, etc., but we did not bother much about them. Since there was a threat to the institution, police protection was sought for a few days, but then, it was a very minor incident.'

However, she did point out that over the last twenty years or so of her career, she has never encountered such an accusation. 'It was strange. I have never felt like that while reporting over all these years, despite covering so many sensitive issues in the past. This was a new experience when our professional ethics were being questioned because of our background. But then, we are all professionals here and have been taught to work in the most unbiased manner possible,' Jaya Menon says.

While Kerala-based journalists living and working in TN did experience hostility while reporting on the dam-related violence and the court proceedings, the scribes who actually faced threats from angry mobs and agitators were the employees of Tamil dailies with editions in Kerala and those of Malayalam dailies with editions in TN.

Rajan (name changed), a senior correspondent working in the Coimbatore bureau of a leading Malayalam daily, recounts the experience of working in Coimbatore when the Mullaperiyar dam row was at its peak as harrowing. 'Although I belong to Kerala, I have been living and working in Coimbatore for over fifteen years now and this is my home. But when the dam row was at its peak, we were looked at with suspicion even by our friends in other media,' he says.

A large team of police personnel were posted outside their office to provide security to the staff and machinery outside their Coimbatore office, and the reporters were generally reluctant to get on with their usual activities. 'But we continued our work and reported exactly what happened here. Our target audience is the Malayalis in Coimbatore and we wrote about their plight. In a sense, we contributed to highlighting the plight of Keralites in Tamil Nadu, which I suppose helped in the state government taking a softer stand. So many Keralites have investments and businesses in Tamil Nadu, especially Coimbatore. When their plights were heard, the politicians started to soften their stand,' Rajan says.

*

In January 2011, a documentary film was released in Kochi by former Supreme Court judge V.R. Krishna Iyer. The film titled *Dams: The Lethal Water Bombs* directed by Sohan Roy was a documentary on the Mullaperiyar dam and was intended to be a 'visual support' to Kerala's claim for a new dam.

The twenty-one-minute documentary that has reportedly won several international awards for best documentary and best editing unfortunately projects only one side of the story, delving into the age of the dam and the possibility of a failure due to the age and weakness of the construction.

At the time of the release, director Sohan Roy had said that it was a thought-provoking film and a curtain-raiser to Dam 999 the first Hollywood venture by a group of Keralites. '*Dams: The Lethal*

Water Bombs is themed on the background of the Mullaperiyar dam. The documentary provides scientific explanation on the disastrous nature of the Mullaperiyar dam, and the work is also a warning to those who turn a blind eye to the reality of a possible disaster,' he said.

The movie that delves mainly into safety concerns of the people regarding Mullaperiyar also raises questions about large dams and their life spans, and ends with strong statements that have enough content to rattle the already petrified population living downstream the Periyar.

The movie ends on the following note: 'Let Mullaperiyar be an eye-opener before it is too late. Yes, today or tomorrow, the Mullaperiyar may prove to be the worst ever watery grave in human history. Let it not happen!' Media reports published at the time of the film's release listed the movie as a visual support for Kerala's claim for a new dam.

While the film received widespread accolades in Kerala and the rest of the world, politicians and farmers in TN saw it as yet another ploy by Kerala to sell lies about the dam at a new forum, that is, through the medium of cinema. MDMK chief Vaiko claims that the movie was sponsored by the LDF government headed by former chief minister V.S. Achuthananthan. 'In fact, the state government had made over five lakh copies of the movie and distributed it freely to everyone concerned with the dam to create a sense of mass hysteria. If it is a film made by an ordinary NRI Malayali, why should the state government take up the job of publicizing the movie?' he asks.

As a response to that documentary film, members of the TN Senior PWD Engineers Association made another documentary film that highlighted the safety features of the dam and was intended to support the state's claim that the Mullaperiyar dam was among the strongest in the country.

This film titled *Mullai Periyar: Problems and Solutions* runs for a length of 42.03 minutes and mainly focuses on the strengthening measures taken by the TN PWD over the last thirty years. In its opening scene, the film claims that the intention of the film is to highlight the real truth about the dam to Tamil people.

'To counter the false propaganda of the Keralites, we too printed several thousand CDs of this documentary film, and went from village to village along the Periyar–Vaigai basin (around 500-odd villages) to distribute the CDs to the people and help them understand the real thought,' Vaiko says.

Retired PWD chief engineer R.V.S. Vijayakumar, claimed that a small group of friends had invested ₹1 lakh from their pockets to make the film. 'Unlike the Kerala government, our politicians do nothing about the issue; hence, we took the initiative and produced the film for reaching the truth to the public. Anyone can see or download it from YouTube,' he said.

*

In August 2012, as I was working on the final drafts of the book, another dam disaster was in the making in neighbouring Kerala, unknown to many outside its immediate impact zone. The Pazhassi dam situated at around 60 kilometres away from Kannur in northern Kerala was overflowing, following heavy rains in Kannur and Kozhikode over the past several days as the southwest monsoon gained more and more strength in the region. As the water level in the dam rose beyond its FRL, engineers at the dam site attempted to open the spillways, but were unsuccessful as they got stuck and water flowed over the dam for several hours before the situation could be brought under control.

According to media reports, the water level in the Pazhassi reservoir exceeded 28.35 metres, which was the road level over the barrage, for the first time in the last thirty-three years. 'Normally, the water in the reservoir is maintained at the full reservoir level (FRL) of 26.52 metres if the shutters could be closed in early November. But the flash floods and landslips caused by heavy downpour in the eastern hill areas bordering Karnataka took the project officials unawares,' said a report in *The Hindu*.

'The dam site was the destination of a large number of people who defied the heavy downpour to witness the flood water overflowing the barrage for the first time since the project's partial commissioning by the then Prime Minister Morarji Desai in 1979. The failure of the irrigation engineers to lift some of its shutters triggered concern about the capacity of the irrigation dam to survive the floods. The 245-metre long barrage, however, is still standing as the water level in the reservoir came down to the level of 18.2 metres on Wednesday from 29.2 metres, that caused fears of an imminent collapse of the dam,' the report said.

While the Pazhassi dam is obviously not as large as the Mullaperiyar dam, the incident that took place is exactly what has been feared by the public living in Idukki district. Whenever a dam overflows and spillways cannot be opened, it is an impending dam disaster. Fortunately, nothing happened and the water level reduced on its own.

Yet, there has not been a fierce discussion or debate on the failure of the engineers in maintaining the dam, and the cause for the spillways not opening has not been debated in a public forum either.

A casual search on the Internet would unveil thousands of blogs and websites containing information on the Mullaperiyar dam. While much debating has taken place in the online world over the Mullaperiyar dam safety, there are hardly any blogs or web posts on the Pazhassi dam overflow or on the condition of dozens of other dams in the same region, that are in worse conditions both in TN and Kerala.

The huge media attention, though, has only been good for the 119-year-old dam. According to a release issued by the National Alliance of People's Movements, a fair number of India's dams are over a hundred years old. The list compiled by the CWC shows there are at least 114 dams in this category, and there are roughly 400 dams which are fifty to a hundred years old.

According to the Madhya Pradesh government, the state has 168 dams which can be called 'distressed dams', out of which sixty-three are less than fifty years old. Since 1917, at least twenty-nine dams have reportedly been damaged. But no state government seem to be as obsessed about the safety of these dams as the governments of TN and Kerala are with the Mullaperiyar.

The huge media attention that has been showered on the Mullaperiyar has probably ensured that this dam is probably the most pampered large dam in India, with TN and Kerala working overtime to prove the other wrong.

*

9

The Home Run

A s I come to the end of this fascinating journey of attempting to unravel the mystery behind a century-old controversy, I still feel like the city-bred novice reporter who has just rushed to cover a riot without knowing what it was all about.

Most of my sources have now stopped talking to me. People who were so enthusiastic in taking me around and explaining the complexities of the conflict now either cut my phone calls within a few seconds of interaction or do not pick up my calls at all.

I went to Kerala as a Malayali with the umbilical cord connection that made be a homeboy in God's Own Country. I was welcomed, allowed to take a peep into the lives of the thousands who live downstream the Periyar and who could get washed away if there was ever a dam break. They shared their fears and anxieties and helped me understand why they feel insecure living beneath an ageing dam.

During my travels among the farmers of Periyar–Vaigai basin and the TN PWD engineers who have been maintaining the dam for so many decades, I went as a Chennai boy, born and raised in the capital city of TN, trying to understand the plight of farmers who in some way make sure that our state granary remains filled. The farmers, most of whom were in the autumn of their lives, shared anecdotes of a life they cherished for over a century and a lifestyle that they fear might get wiped away from the face of the earth if the existing dam was decommissioned.

As I began asking difficult questions towards the end of my journey, the same people who welcomed me earlier started fearing that I was an agent of the enemy trying to take away a cause that they had fought for years together, a cause that has now become much more than a mere dam controversy.

Yes, the people of Kerala need a safe dam and a right to live peacefully without any fear. In the same vein, the thousands of farmers in TN do need water to carry on with their lives without being subjected to poverty and displacement.

After having spent close to a year tracking down every possible detail about the 119-year-old dam and the controversy surrounding it, I am still uncertain if there is a possible solution that is acceptable to both the states.

If one were to believe that the SC could solve the Mullaperiyar dam row, then the hundred-year-old dispute would be solved within a year, two at most. The SC appointed committee has already come out with a report on the safety of the dam which will be debated over the next few months, and a judgment would also arrive in due time. Unfortunately, there appear to be no takers for their belief. Neither the people of Kerala nor TN believe that the Mullaperiyar dam row could be resolved anytime soon, or could even be solved at all.

But then, this book was not meant to be a solution for any problem in the first place. It was neither meant for the protestors in Chappathu, who have been on a strike for six years now, nor the farmers in the Periyar–Vaigai basin, who have been fighting tooth and nail at every available forum to protect their rights to the water. All these people, along with irrigation department engineers in Kerala and the PWD engineers in TN, know a lot more than I do and have been tracking this issue all their adult lives.

It was written for the lakhs of urban Indians like me who are seldom aware of the complex livelihood concerns of people who live far away from the bustling metros of India, citizens who only see the Mullaperiyar dam controversy, or the Cauvery dispute or any other such issue, as a mere irritant to their luxurious way of life, an irritant that they wish would vanish from the face of this planet through the waving of some magic wand.

This is a mere attempt at understanding a complex problem, and as they say, knowing and accepting that we have a problem is the first step towards resolving the problem. If the software engineers, BPO workers, the high-end businessmen and the Gulf Malayalis and Tamil Americans understand exactly what is going on with the Mullaperiyar dam, perhaps there would be a political solution possible that does not hurt the sentiments of both Malayalis and Tamils.

*

One of the first persons I had interviewed in connection with this book is Professor A. Mohanakrishnan, a water management expert who had spent his entire life working on dams and irrigation systems, and an engineer, respected by his peers from across the country. Even after more than two decades after his retirement from active service, Mohanakrishnan still visits his modest office inside the campus of the Government Museum in Chennai where he sits as the head of the Cauvery Technical Cell, the wing that is fighting a similar decades-old water dispute with Karnataka.

Mohanakrishnan despises politics and politicians. 'If you want to talk anything about the current controversy, please leave this place right away,' he told me. 'If you want to talk about the dam, the marvel of its engineering and its utility, we can talk,' the professor said.

'Pennycuick had achieved whatever he could because he had a sense of purpose. He wanted to contribute positively to the society and was driven by a deep sense of commitment. At the time he conceived the project, it was the first inter-basin transfer attempted anywhere in the world. It takes a vision to construct something like this,' he said, exposing his awe of the Mullaperiyar dam even so many years later.

After lengthy conversations with Mohanakrishnan on dams and their benefits, I finally asked him what he thought about the Mullaperiyar and its lifespan.

'I believe water should be utilized for whoever needs it most. I do not care whether it is the people of Kerala or Tamil Nadu but water should be used most effectively, and as engineers it is our job to ensure that,' he said. 'The Mullaperiyar waters were otherwise useless at the time the dam was built. Even now, one should find out who needs it most and give it to them.'

When asked if there was a lifespan for a dam as claimed by many, the veteran water expert said that he believed that technically, there was no limit for the life of a dam. 'It will be just as good as long as you maintain it well. But then, rivers bring along with them a lot of sediments that get deposited at the bed of the reservoir. Over a period of time, there would be so much silt accumulated at the bottom that storage levels would drastically reduce and the dam would no longer be useful. I believe that is when a dam becomes useless to anyone,' he said. 'Until then, water should be shared to generate the best benefit of the water in the right way. But that is my personal opinion,' he added.

*

Senior engineer James Wilson works with the Kerala State Electricity Board and is presently a member of the Mullaperiyar Cell, Government of Kerala. Although there are several engineers who are instrumental in providing technical details for the advocates appointed by the Kerala state to fight the case in the SC, James is one of the most resourceful and has been working on the issue for several years now. He is considered an expert on the Mullaperiyar in his state.

James believes that the dam is hydrologically and seismically unsafe and decommissioning it could be the best option available. 'The Mullaperiyar dam was built at a time when dam engineering was at its infancy and safety standards were not as good as now. Today, basic engineering mathematics can prove that the dam is unsafe. We do not have to go by hearsay anymore,' he says.

'Experts from IIT Roorkee and IIT Delhi have conducted technical studies and have found the dam to be unsafe. We will contest in the court too and prove our point,' says Mr Wilson. 'Our claim is based on the Precautionary Principle, that is, if there is any possibility that there could a flood that is larger than the maximum predicted flood or if there is an earthquake that could rattle the structure, then we have to apply this principle and decommission the dam. The safety of a dam is the most important factor when you consider whether to decommission or rebuild it,' he says.

James Wilson believes that people should set aside politics and controversies and allow the engineers to sort out the issue peacefully. 'It is an engineering problem and we should solve it. If you remove politics and other controversies from the dispute, the Mullaperiyar problem can be solved in a matter of days,' he says.

*

In the most recent report on the safety standards of the Mullaperiyar dam filed by the EC appointed by the SC, the technical and legal experts, while affirming that the dam was completely safe for the known future if well maintained and periodically strengthened, also made a special mention on the way forward and suggested two alternative approaches towards a long-term solution to the problem.

The first alternative was the construction of a new dam by the state of Kerala, at its own expense, to serve to allay the fears of the people of Kerala, if techno-economically cleared by the planning commission and the ministry of environment and forests. This dam would be constructed even as the existing dam remains in operation, and would be maintained by an independent committee comprising of members from TN and Kerala. The court also pointed out that the access of TN's rights to the Periyar waters should not be compromised.

The second alternative for a long-term solution, as suggested by the EC, was the construction of a tunnel at a height of 50 feet that would be able to drain the reservoir to that level, so that repair works could be done whenever needed without much effort. The existing tunnel of the dam is at a height of 106.5 feet up to which level only can the reservoir be drained. The new tunnel would also ensure that flood water could be drained faster in the event of floods, thus decreasing the hydrological safety concerns of the people of Kerala.

While this alternative has been suggested by the courts now, the absence of a tunnel at 50 feet or so to drain the reservoir for repairs was cited as a deficiency in the construction as early as the 1890s, when Pennycuick made a presentation to the Royal Society of Engineers on the Periyar Project. Some of the experts gathered at the meeting pointed out that the dam ought to have had a tunnel to drain the reservoir completely. This design flaw that was overlooked for the last 120 years has now been suggested by the EC as a long-term alternative to ensure that the dam remain safe in the event of floods, and that repair work can be done easily.

*

Earlier in the book, I had included chapters on the history of the construction of the Periyar dam 120 years ago, and the technical aspects of the safety of a new dam according to modern standards. To the casual reader who tries to understand the controversy issue, these two chapters might seem unnecessary. But they had been included mainly for the reader to understand the magnitude of the task of constructing a dam.

The tonnes of cement, iron, rocks and other construction material that need to be transported to the dam site, the thousands of construction workers who need to be transported to the site, and who need to have a place to live and work, the roadways needed for the giant earthmovers and cranes to be deployed in the construction, the blasts that need to take place to prepare the river bed for the dam, the hospitals and kitchens that need to be set up to feed and take care of the workers, besides other peripheral activities such as carpentry, electrical works, etc.—despite having advanced dam construction techniques, structural engineers highlight that constructing a dam could take several years and may be completed only in stages over the years.

The Mullaperiyar dam is located in the core area of the Periyar Tiger Reserve, one of the most protected forests in the country. Stretching across 925 square kilometres of pristine forests that thousands of species of birds, bees, butterflies, animals and fish call home, the Periyar Tiger Reserve is one of twenty-seven tiger reserves in India and is fiercely guarded by the Kerala forest department.

Considering the importance of this sanctuary with respect to its tiger population, the area was brought under Project Tiger in 1978 as the 10th Tiger Reserve in India, named as Periyar Tiger Reserve (PTR). In 1982, an extent of 350.54 square kilometres was notified as Periyar National Park. In 1991, the area was included as part of Project Elephant Reserve No. 10.

According to a recent census, at least twenty-four of the remaining 1,700-odd tigers surviving in the world call this reserve their home. Besides, several other species of animals classified under the Wildlife Protection Act are found in these forests.

Back in 1887, Lord Connemera, then governor of Madras, inaugurated the Periyar dam project by felling a tree at the site of the dam. Today, if Lord Connemera himself would come for the inauguration of the construction of a new dam and fell a tree inside the Periyar Tiger Reserve, he could be arrested. Cutting of trees is illegal inside the core areas of any tiger reserve in the country.

The Ministry of Environment and Forests has laid down extremely strict guidelines for any kind of development or construction activity in any of the reserve forests in the country. Tiger reserves are the most protected among them.

The SC is also in favour of protecting the last stretches of forests that remain for the benefit of the tigers, elephants and millions of other species of animals, birds and insects that have been pushed to the brink of extinction in the name of civilization. An interim order issued in August 2012 had even banned tourism in tiger reserves across the country until further direction.

When the British planned to construct a dam across the Periyar, they had other things in mind. While wiping away famine in the Madurai region was a noble cause for which the dam was built, this noble effort was delayed by over thirty years mostly because the engineers who proposed the dam could not show a business plan that gave the British 6 per cent profit on the investment. Besides being a water sources the Mullaperiyar dam was mostly a sensible commercial decision that did not take into account the ecological damage that could have arisen by diverting a west flowing river to the east.

In twenty-first century India, pristine reserve forests are more heavily guarded and protected than big cities. Getting environmental clearance even for laying a road inside a tiger reserve might be a task that could be considered next to impossible. In such a forest, getting environmental clearance for a ₹1,000 crore dam project by clearing acres and acres of forest land for the sake of a reservoir, thus displacing thousands of species of animals, birds and insects, could actually be considered impossible.

Even if the MoEF could be coerced into accepting the proposal, Kerala is known across the country for its environmental activists. Even as early as 1975, a strong movement organized by environmentalists against a proposed dam on the Kunti River inside the Silent Valley Reserve forests actually forced the then Prime Minister Indira Gandhi to abandon the project to protect the rainforests.

To construct a dam during the 2010s in a land that is famous for protecting a forest more than thirty years ago, at a time when forest cover was more extensive, could be an arduous task for the state government. The legal battle for getting environmental clearance for a new project inside the Periyar Tiger Reserve could take several more decades.

*

In many ways, the problems involved in solving the Mullaperiyar dam row is a good metaphor for most issues that bleed this country. The politicians do not listen to the technical experts, the technical experts do not listen to the needs of the common people, and the people have learnt over time never to trust their politicians.

As the two states continue to fight it out to make the Periyar flow their way, much silt would have accumulated in the Periyar reservoir to make it not worth a fight.

*

10
The Bigger Picture

Wtraces of water on the surface of the moon in September
2009, scientists across the world were jubilant. For to
find water is to find life.

Our history is replete with anecdotes of rivers mothering
civilizations along their banks. From the Nile that gave birth
to ancient Egyptian civilization to the Euphrates and Tigris
responsible for the Mesopotamian civilization to our own Indus
Valley Civilization, rivers have always been associated with
abundance of life, prosperity and growth.

For centuries, our ancestors have practised the art of storing
and diverting river water into barren lands to raise crops, using
techniques available to them. The fact that agricultural produce
still forms a significant part of our GDP vindicates the importance
of water in our country.

Evidence of water management in India has been traceable
throughout our long history. Although the pattern of rainfall in
the country is highly seasonal, with 50 per cent of the precipitation
falling in just fifteen days, and a majority of river flows occurring
in just four months, the people have adapted to this variability
by either living along riverbanks or by careful harvesting and
management of water.

As early as the fifth century AD thousands of minor irrigation
tanks had been constructed by the Cheras, Cholas and Pandyas

to store water and divert them for agriculture. Most of this management was at the community level, relying on imaginative and effective methods of harvesting water in tanks and small underground storages.

Small storage reservoirs were constructed before the Mauryan era around 300 BC and the Grand Anicut across the Cauvery was built in the second century. During the Mughal era (tenth through nineteenth centuries) large-scale run of the river schemes and inundation canals were constructed. As the population of the country grew, so did the demand for water and the need for effective management of this valuable resource.

According to a report compiled by the World Bank on the country's water management scenario, India currently has approximately 17 per cent of the world's population, but shares only 4 per cent of the country's water resources and 2.6 per cent of total land available. The report titled 'India's Water Economy: Bracing for a Turbulent Future' projects a bleak picture of the water scenario in the country and a turbulent future ahead. It states that the current water development and management system is not sustainable, unless dramatic changes are made soon in the way the government manages water. India will neither have the cash to maintain and build new infrastructure nor the water required for the economy and the people.

Despite taking up large dam projects, the country can still store only relatively small quantities of its fickle rainfall. While more arid but rich countries such as the United States and Australia have built over 5,000 cubic metres of water storage per capita, and even middle-income countries like South Africa, Mexico, Morocco and China can store about 1,000 cubic metres per capita, India's dams can store only 200 cubic metres per person. We can store only about thirty days of rainfall when compared to 900 days in major river basins in arid areas of developed countries. It is apparent that there is every indication that the need for storage would grow because global climate change is going to have major impacts in India— there is likely to be rapid glacial melting in the coming decades in the western Himalayas, and increased variability of rainfall in large parts of the subcontinent, as predicted by several climate change studies.

A review of India's hydro power infrastructure also reveals a dark picture. While industrialized countries harness over 80 per cent of their economically viable hydro power potential, in India the figure is only 20 per cent, despite the fact that the Indian electricity system is in desperate need of peaking power and that Himalayan hydro power sites are, from social and environmental perspectives, among the most benign in the world.

World Bank water management experts point out that especially in the water-rich northeast of the country, water can be transformed from a curse to a blessing only if major investments are made in water infrastructure (in conjunction with 'soft' adaptive measures for living more intelligently with floods). In many parts of the country there are also substantial returns from investments in smaller-scale, community-level water storage infrastructure (such as tanks, check dams and local water recharge systems). And there is a massive need for investment in water supply systems for growing cities and for underserved rural populations.

The National Water Commission of 1999 has shown that overall water balances are precarious and that crisis situations already exist in a number of basins. By 2050, demands will exceed all available sources of supply as already around 20 per cent of the aquifers are in a critical condition, a number set to grow to 60 per cent in the next twenty-five years unless there is a change. While regulations have been put in place across most of the country regarding bore wells, a large amount of ground water is still being exploited and misused, adding to the water woes of the nation.

The draft of the National Water Policy (2012) recommended by the National Water Board admits that large parts of India have already become water stressed and that rapid growth in demand for water, due to population growth, urbanization and changing lifestyle, poses serious challenges to water security. The policy draft also raises serious concerns about inter-regional, interstate, intra-state, as also inter-sectoral disputes in sharing of water, which strain relationships between neighbouring states and hamper the optimal utilization of water through scientific planning on basin/ sub-basin basis.

*

For two countries that have waged three wars with each other in the past and are always at loggerheads with each other, the sharing of water between India and Pakistan has been exemplary and has never led to a full-fledged skirmish until recently.

The Indus and its tributaries originate in the Himalayas and flow through the plains of Punjab and Sind before draining into the Arabian Sea. According to World Bank estimates, approximately 26 million hectares of land is irrigated by this system, considered the largest irrigation network in the world. The Indus Water Treaty signed fifty years ago is still seen as the best example of interstate cooperation between two countries in sharing common rivers.

The Indus system of rivers consists of six major rivers. While three of them—Ravi, Beas and Sutlej—flow eastwards, the other three—Indus, Jhelum and Chenab—flow west. In a broad sense, the Indus Water Treaty guarantees India all the exclusive rights of the east-flowing rivers before the point when they enter Pakistan, while the neighbouring country gets exclusive rights of the west flowing rivers. This treaty moderated by the World Bank was signed in September 1960 and is still considered one of the best water-sharing agreements.

Over the last sixty years since Independence, the demand for power has been surging in the country, and the deficit between demand and supply has been widening. With a looming power crisis, the Indian government has planned for over thirty hydro-electric projects on the Indus and its tributaries in Jammu and Kashmir and elsewhere to generate electricity. This move has been stiffly opposed by the Pakistani government who claim that building such hydro projects would drain the river and thereby leave very little water for the Pakistani farmers, who are already suffering from water shortage.

One such flashpoint is the Kishenganga hydroelectric project proposed to be built on one of the tributaries of Indus in Jammu and Kashmir. The dam built by India will divert a portion of the river down a 22-kilometre long tunnel into turbines for generating electricity. Even as the efforts are underway for the construction of this dam, Pakistan is building another dam on the same river downstream (called Neelum in Pakistan) with an installed capacity of 960 MW of power. Pakistanis fear that the dam built by India would restrict its generation capacity by half, and are putting up stiff resistance to the Indian project.

As mentioned in the Indus Water Treaty, the matter has been referred for arbitration by the International Court of Arbitration (ICA), where a committee consisting of representatives from India and Pakistan as well as international members will decide on the future of these dam projects.

The other project that had potential for conflict between India and Pakistan is the Baglihar dam, a run of the river project on the Chenab in the southern Doda district of Jammu and Kashmir, with a capacity for generating 900 MW of electricity. While Indians claimed that the dam did nothing to the water flowing into Pakistan except causing a minor delay to generate hydroelectric power from it, Pakistanis see it as a means for India to control its water and, thus, put lives of lakhs of farmers living downstream in peril.

When Pakistan objected to the dam's design, India accepted international arbitration, which was the first time that option was invoked in the Indus Water Treaty. International experts studied the dam and suggested minor changes to the design, before concluding that it posed no threat to Pakistan's right to access the Chenab waters. After years of negotiation and arbitration, in September 2010, both the countries resolved to accept the verdict of the ICA and not to raise the issue any further.

But the water woes between India and Pakistan are far from over. With India planning at least thirty more hydroelectric projects to cater to the surging energy demands of the country, Pakistan, which is downstream, is always sceptical of the motives of the Indian government and sees these developments as a means for the larger nation to control access to their waters.

*

While India has the advantage of being the upstream user of the Indus in the west, the situation is exactly the opposite in the eastern part of the country where it shares the mighty Brahmaputra with China and Bangladesh.

For years, the Indian government has been worried about the Chinese government building a dam on the Brahmaputra (called Tsangpo in China). The Brahmaputra originates in the Tibetan plateau and flows into India through Assam, providing drinking water and irrigation for much of northeastern India, before joining the Teesta River to enter Bangladesh.

Despite repeated assurances by the Chinese government that the construction of a dam over the Tsangpo were all run of the river projects and would not affect the river's flow in India, the Chinese statements have been received with scepticism, and there is a strong lobby in India that fears that China will divert the Brahmaputra waters into their territory to irrigate lakhs of acres of farmlands in central China.

Although several proposals have been mooted by Chinese engineers in the past to divert water from the water-abundant southern part of China to the drier, more prosperous north by building dams on the Tsangpo and other rivers, these plans appear to be technically impossible as of now, according to experts in the field.

In 2010, the Chinese confirmed that they were building a dam (Zangmu Dam) on the Brahmaputra but assured that it would not have any impact to the course of the river flowing downstream. At the moment, the construction of the dam is underway and is expected to be completed by 2015, with an installed capacity of generating over 500 MW of electricity.

The Chinese assertion does not allay India's fears one bit simply because it is a downstream country for this river. However, the country's water-sharing agreements with the other eastern neighbour Bangladesh seems to be booming, with the country planning several dams in the rivers flowing east.

*

A report released by the US Intelligence Community on Global Water Security in March 2012 has suggested that the risk of water wars would increase substantially ten years from now, with global water demand likely to outstrip current sustainable supplies by 40 per cent by 2030.

The report also forecasted that while, historically, water tensions have led to more water-sharing agreements than violent conflicts, once there is not enough water to go around, these fragile pacts may collapse, with more powerful upstream nations impeding or cutting off downstream flow.

It is time we look at our rivers and dams differently.

*

Epilogue

May 2014: Supreme Court Delivers Verdict on the Mullaperiyar Dam

Almost two years after the Supreme Court-appointed Empowered Committee submitted its report on the safety of the Mullaperiyar Dam, after a detailed study of the 119-year-old dam's safety standards, the SC on 7 May 2014 delivered its verdict in the case.

The Constitution Bench of the Supreme Court, comprising Chief Justice R.M. Lodha, Justice H.L. Dattu, Justice Chandramauli K.R. Prasad, Justice Madan B. Lokur and Justice M.Y. Iqbal, upheld the rights of TN to raise the level of the Mullaperiyar Dam from 136 feet to 142 feet and subsequently to 152 feet after all the strengthening measures were completed. The bench also pronounced that the Kerala Irrigation and Water Conservation (Amendment) Act 2006 was unconstitutional and could not stand in the way of the apex court's 2006 verdict.

The court also came down heavily on the state of Kerala for obstructing dam repair works in the past and stated that Kerala could not obstruct TN from raising the water level of Mullaperiyar dam to 142 feet and prevent it from carrying out repair works as per its judgement dated 27 February 2006.

In an effort to allay the apprehensions of Kerala about the safety of the Mullaperiyar dam on the restoration of FRL to 142 feet, the court has directed the appointment of a three-member Supervisory Committee. The committee would have one representative from the Central Water Commission and one representative from each of the two states—TN and Kerala. The representative of the Central Water Commission would be its chairman.

As per the SC verdict, the powers and functions of the Supervisory Committee shall be as follows:

(i)   The committee shall supervise the restoration of FRL in the Mullaperiyar dam to the elevation of 142 ft.

(ii) The committee shall inspect the dam periodically, more particularly, immediately before the monsoon and during the monsoon and keep close watch on its safety and recommend measures which are necessary. Such measures shall be carried out by Tamil Nadu.

(iii) The committee shall be free to take appropriate steps and issue necessary directions to the two states—Tamil Nadu and Kerala—or any of them, if so required, for the safety of the Mullaperiyar dam in an emergent situation. Such directions shall be obeyed by all concerned.

(iv) The committee shall permit Tamil Nadu to carry out further precautionary measures that may become necessary upon its periodic inspection of the dam in accordance with the guidelines of the Central Water Commission and Dam Safety Organization.

With the most recent verdict, the Mullaperiyar dam conflict returns to where it was eight years ago, following the previous SC verdict. While the TN government is busy making preparations to gradually raise the water level in the dam to 142 feet, bringing cheer to lakhs of farmers in the Periyar–Vaigai basin, the Kerala government is making plans to file a review petition before the SC over the next few months in an obvious effort to appease the residents of Idukki district, who are distraught by the verdict.

*

Acknowledgements

Writing a book on any controversial topic is difficult. It is especially arduous when the controversy concerns two neighbouring states engaged in a pitched battle, both of which have a deep-rooted connection with the author.

Despite the intricacies involved, this work was mainly possible because of my agent Kanishka Gupta of Writer's Side who first mooted the idea for this book and pushed me till I completed the manuscript.

I thank my bosses at *Deccan Chronicle* who gave me the freedom to take time off work whenever I wanted to and do leg work for the book and the retired and serving engineers at the Tamil Nadu Public Works Department, the Kerala State Electricity Board and the Mullaperiyar Cell of the Kerala government who spent several hours of their time to make me understand the complexity of dam engineering and its safety concerns. I also have to thank all those people who live on either side of the Mullaperiyar dam between Madurai in TN and Kottayam in Kerala, who welcomed me to their homes and spoke at length about their woes and enlightened me on the issues they face.

Most of all, I thank my family and friends for supporting me through this endeavour and for putting up with my eccentricities.

References

A.T. Mackenzie (1899). *History of the Periyar Project*, Madras: Government Press.

Col John Pennycuick (1897). *The Diversion of the Periyar*, United Kingdom: Proceedings of the Institution of Civil Engineers, Journal of the Western Society of Engineers.

J.W. Barry, Sir R. Sankey, G. Farren, E.P. Hill, L.F.V. Harcourt, G.J. Symons, G.F. Deacon, W.H. Hall, A.R. Binnie, F.J. Waring, A. Chatterton, W.C. Unwin and J. Pennycuick (1897). *Discussion. The Diversion of the Periyar and the Periyar Tunnel*, United Kingdom: Proceedings of the Institution of Civil Engineers, Journal of the Western Society of Engineers.

Prof. A. Mohanakrishnan (1997). *History of the Periyar Dam with Century Long Performance*, New Delhi: Central Board of Irrigation and Power.

Er R.V.S. Vijayakumar and Er A. Veerappan (2012). *Save Mullaiperiyar Dam*, Chennai: Tamil Nadu Public Works Department Senior Engineers' Association.

Prof. D.K. Paul (May 2008). *Structural Stability of Mullaperiyar Dam Considering the Seismic Effects*, Roorkee: Department of Earthquake Engineering, Indian Institute of Technology, Roorkee.

Prof. D.K. Paul (2009). *Seismic Stability of Mullaperiyar Composite Dam*, Roorkee: Department of Earthquake Engineering, Indian Institute of Technology, Roorkee.

Madhusoodhanan C.G. and Sreeja K.G. (2010). *The Mullaperiyar Conflict*, Bangalore: National Institute of Advanced Studies, Indian

Institute of Science Campus.

A.K. Ghosain, Subash Chander (2008). *Probable Maximum Flood Estimation and Flood Routing Study for the Mullaperiyar Dam* (2008), New Delhi: Civil Engineering Department, Indian Institute of Technology, Delhi.

Justice (Dr) A.S. Anand, Justice K.T. Thomas, Justice (Dr) A.R. Lakshmanan, Dr C.D. Thatte, D.K. Mehta (2012). *Report of the Empowered Committee on Mullai Periyar Dam Dispute Issue Between the States of Tamil Nadu and Kerala.*

The Dam Safety Bill (2010). Lok Sabha

David S. Bowles, Francisco L. Guiliani, Desmond N.D. Hartford, J.P.F.M. (Hans) Janssen, Shane McGrath, Michel Poupart, David Stewart, Przemyslaw A. Zielinski (2007). *ICOLD Bulletin on Dam Safety Management*, International Commission on Large Dams.

Water Resources Law, Berlin Conference (2004). International Law Association

Salman M. Salman (2007). *The Helsinki Rules, the UN Watercourses Convention and the Berlin Rules*. World Bank, Washington: Water Resources Development, Vol 23, No. 4 (December 2007).

World Bank Report No. 34750-IN (2005). *India's Water Economy: Bracing for a Turbulent Future* (22 December 2005)

Central Water Commission (2009). *National Register for Large Dams – 2009*, New Delhi

Central Water Commission (1986). *Report on Dam Safety Procedures*. New Delhi: Dam Safety Organization, Central Water Commission, Ministry of Water Resources.

Dam Safety Organization (1987). *Guidelines for Safety Inspection of Dams*. New Delhi: Central Water Commission, Ministry of Water Resources (June 1987).

Dilip D'Souza (2002). *The Narmada Dammed: An Inquiry into the Politics of Development*, Penguin Books, Navi Mumbai.

Patrick McCully (2001). *Silenced Rivers: The Ecology and Politics of Large Dams*, London, New York: Zed Books

'Indian scientists rejoice as Chandrayaan – 1 Traces Water on Moon' (2009), *The Times of India*, New Delhi, 24 September.

Water Security for India: The External Dynamics (2010). Institute for Defence Studies and Analyses Task Force Report, New Delhi, Institute for Defence Studies and Analyses, September 2010.

Steven Lee Myers (2012). 'US Intelligence Report warns of global water tensions', *The New York Times*, 22 March.

'Rain blitz weighs down Pazhassi dam' (2012), *The Hindu*, Mattannur, 8 August.

'Pazhassi dam is now safe' (2012), *The Hindu*, Kannur, 8 August.

Current Interstate Water Disputes and Tribunals, Ministry of Water Resources, from http://wrmin.nic.in

Periyar Tiger Reserve History from http://periyartigerreserve.org

Interviews

Former Kerala minister and Revolutionary Socialist Party (RSP) leader N.K. Premachandran, Kollam, May 2012.

Fr Joy Nirappel, Mullaperiyar Samara Samiti, Uppukara, Idukki district, May 2012.

Kerala State Electricity Board Senior Engineer and Member, Mullaperiyar Cell James Wilson, Kollam and Trivandrum, May 2012.

Thomas Mathew, resident of Vallakadavu and campaigner for new dam, Vallakadavu, Idukki district, May 2012.

Joseph Karoor of Mullaperiyar Environment Protection Forum, Kumily, May 2012.

Editor of *Tughlaq* magazine and leading political analyst in TN Cho Ramaswamy, Chennai, August 2012.

Marumalarchi Dravida Munnetra Kazhagam (MDMK) Chief Vaiko, Coimbatore, August 2012.

R.V.S. Vijayakumar, former chief engineer (Madurai), Tamil Nadu Public Works Department, Madurai, May 2012.

Former president of Periyar-Vaigai Farmers' Welfare Association

K.M. Abbas, Cumbum, May 2012.

Ramasamy, Periyar-Vaigai Farmers' Welfare Association, Madurai, May 2012.

Prof. A. Mohanakrishnan, one of the foremost experts in dams in TN and author of *History of the Periyar Dam With Century Long Performance*, Chennai, March 2012.